AIR, WATER AND SOIL POLLUTION SCIENCE AND TECHNOLOGY

SOIL MANAGEMENT

TECHNOLOGICAL SYSTEMS, PRACTICES AND ECOLOGICAL IMPLICATIONS

AIR, WATER AND SOIL POLLUTION SCIENCE AND TECHNOLOGY

Additional books in this series can be found on Nova's website
under the Series tab.

Additional e-books in this series can be found on Nova's website
under the e-book tab.

AIR, WATER AND SOIL POLLUTION SCIENCE AND TECHNOLOGY

SOIL MANAGEMENT

TECHNOLOGICAL SYSTEMS, PRACTICES AND ECOLOGICAL IMPLICATIONS

MANUEL ESTEBAN LUCAS BORJA
EDITOR

nova publishers
New York

Additional color graphics may be available in the e-book version of this book.

Library of Congress Cataloging-in-Publication Data
Soil management : technological systems, practices, and ecological implications / editor, Manuel Esteban Lucas Borja (Universidad de Castilla La Mancha, Escuela Ticnica Superior de Ingenieros Agrsnomos y de Montes, Departamento de Ciencia y Tecnologma Agroforestal y Genitica Campus Universitario, Albacete, Spain).
pages cm. -- (Air, water and soil pollution science and technology) includes index.
ISBN 978-1-63483-274-8 (hardcover)
1. Soil management. 2. Soil ecology. I. Lucas-Borja, Manuel Esteban
S591.S6956 2015
631.4--dc23

Published by Nova Science Publishers, Inc. † New York

CONTENTS

PREFACE

Natural ecosystems provide the basic conditions without which humanity could not survive. Goods and services provided by ecosystems include, for example, provisions of food, fibre and fuel, purification of water and air, cultural and aesthetic benefits, stabilization and moderation of the Earth's climate, generation and renewal of soil fertility, including nutrient cycling or maintenance of genetic resources as key inputs to crop varieties and livestock breeds, medicines, and other products. However, the ability of natural ecosystems to continue performing these services is seriously threatened since the diversity of plant species and soil are being seriously deteriorated and, in some cases, destroyed. Various studies worldwide have shown that soils do not support intensive annual plant cultivation without fertilizer applications and even these may not maintain sustainability. Inappropriate silvicultural operations (or the use of land for intensive agricultural purposes) is one of the main causes of soil degradation, and there is therefore worldwide interest in quantifying the loss of soil quality generated by incorrect agricultural operations or forest management practises. This can only happen if people have the right information, skills, and organizations for understanding and dealing with soil and plant diversity issues. The book shows different studies and research works on the topic of soil management, such us soil degradation, microbiological soil properties, bio-fertilizer soil applications, water dynamics in soil profile, soil erosion and sustainability of soils among others. Transferring information and knowledge within the society is crucial for fighting soil management and soil deterioration. In addition, promoting the sustainable use of soil quality will be of growing importance for maintaining society and biodiversity in the years and decades to come.

Dr. Manuel Esteban Lucas-Borja
Castilla La Mancha University

In: Soil Management
Editor: Manuel Esteban Lucas Borja

ISBN: 978-1-63483-274-8

Chapter 1

IMPACT OF MANAGEMENT PRACTICES ON SOIL QUALITY: ECOLOGICAL IMPLICATIONS

Maria Cristina Moscatelli, Sara Marinari and Ruxandra Papp
Dept. for Innovation in Biological, Agrofood and Forest systems, University of Tuscia, Viterbo, Italy

ABSTRACT

Soil management concerns all operations, practices and treatments used to protect soil and enhance its performance. These practices may be different depending on the specific land use, ranging from tillage and fertilization in agricultural soils to grazing and harvesting in pastures and forests, respectively. Whatever the specific land use may be, any management practice can exert a strong impact on soil enhancing or depleting its quality and, ultimately, its numerous functions. In fact, soil physical, chemical and biological properties may be differently affected by soil management depending on a specific kind of practice or treatment, on soil intrinsic pedological characteristics and on the responses of soil biota. In particular, negative impacts of certain practices may be buffered by the capacity of soil biota to continue to carry on their functions thus reducing these impacts on key ecological processes and properties (e.g., decomposition, nutrient cycling, detoxification, biodiversity etc.).

Soil quality can be assessed by means of physical, chemical and biological indicators whose sensitivity, and consequent reliability, has been widely tested in different land uses and under different management practices.

In rural environments soil quality may be affected by the adoption of different agricultural systems such as organic, biodynamic or conventional agriculture. The effect of various levels of tillage, crop rotations, fertilization etc. has been monitored with respect to carbon (C) sequestration, microbial abundance and diversity, nutrients availability and leaching, soil structure. In particular, indicators focused on soil organic matter quality more than quantity allow better predictions on soil carbon storage capacity.

Forest clearing, coppicing or thinning, fertilization cycles, prescribed fires may produce significant changes in soils of managed forests or plantations that could unbalance nutrient cycling, C emissions and microbial diversity and functions.

Finally, intensive livestock grazing and/or abandonment of traditional practices in pastures and grasslands modify soil capacity to sequester C, aggregation and consequent resistance to water erosion particularly in mountain ecosystems.

Monitoring soil quality in response to management practices is necessary also for the potential synergic effects that may derive from the interaction with other pressures of natural origin such as climate changes.

The aim of this chapter is to review the main effects and ecological implications that different management practices have on key issues relevant for soil functioning and ecosystem services.

Keywords: soil, management practices, soil quality, indicators

INTRODUCTION

Soil functioning is fundamental to virtually all of the ''provisioning'', ''regulating'' and ''cultural'' services identified in the Millennium Ecosystem Assessment (Powlson et al., 2011).

Since soils are a vulnerable, non-renewable natural resource, there is an urgent need to maintain their numerous functions, to sustain biological productivity and to serve as an environmental buffer against disturbances. Soils provide a multitude of ecosystem processes, which have been denoted ecosystem services or life support functions (LSF) (Pereira e Silva et al., 2013). The capacity of soils to sustain such functions is a key feature of what has been denominated "soil health/quality", although there is still an active debate on the differences between these two definitions (Gil Sotres et al., 2005). Utilising soil for agriculture inevitably leads to changes in its properties such as nutrient status, pH, organic matter content and physical characteristics. In many cases changes that are beneficial for food production are detrimental for other ecosystem services, so there is a tension between the different functions of soils (Powlson 2011).

Different land uses and consequent management practices can deeply affect soil quality as well as the sustainability of production systems. Sustainable agriculture aims at meeting the present needs without compromising the productive potential necessary to sustain future generations. Rational soil use practices must allow economically and environmentally sustainable yields, which will only be reached with the maintenance or recovery of the soil resource (Cardoso et al., 2013). The main concerns of land use change, present in the international policy agendas involved in the mitigation of greenhouse gases, are the effects on carbon (C) stocks. Changes in land use and soil management practices induce consistent variations in soil organic C stocks, and are largely responsible for increases in atmospheric CO_2 from terrestrial ecosystem (Guo and Gifford, 2002; Canadell et al., 2000). Furthermore, the change from an ecosystem to another could occur naturally or be the result of anthropic impact such as conversion from natural forest to arable land.

Accordingly, a priority for soil research is to provide the basis for management practices that avoid irreversible damage to the soil resource, leading to agricultural systems that are sustainable soil physical, chemical and biological properties can be severely altered by management practices, in most cases irreversibly.

Monitoring soil quality implies the use of indicators that can help to identify problems, to make estimates of sustainable food production, to monitor changes in environmental quality, to assist government agencies in formulating land - use policies (Doran and Parkin, 1994). Soil quality indicators are commonly divided into physical, chemical and biological indicators (bioindicators), being these last extremely sensitive and responsive to changes. Bioindicators are in fact also used as early warning of degradation conditions and can help to focus stress conditions before an irreversible level is reached. A very critical aspect in monitoring soil quality is in the choice of the correct indicator to be determined. Indeed, it is very important to understand which soil properties can be affected by specific land uses and management practices so to select the correct set of indicators that may inform on soil quality status. Indicators may inform on the impact on soil quality but, at the same time, may point to the resilient capacity that each soil shows when facing disturbances and/or stress conditions induced by mis-management.

Soil organic matter and C storage, nutrient cycling, biological properties and biodiversity, structure and erosion can be considered key issues that influence soil correct functioning within terrestrial ecosystems and in relation to ecosystem services. The following paragraphs will then focus on the effects of soil management on these key soil properties.

Soil Organic Matter – C Cycle

Soil organic matter (SOM) is a key attribute of soil quality and is usually considered as one of the most important properties of soils because of its impact on ecosystem sustainability and its effects on other soil physical, chemical, and biological characteristics (Schmidt et al., 2011). For this reason, as an indicator, it may be considered as a proxy of soil quality in general. One of the most important implications of SOM is its role in storing C into the soil. Of the major global C pools, terrestrial ecosystems hold the potential to capture and store C in SOM through changes in management that are also of benefit to the multitude of ecosystem services that soils provide. This potential can only be realized by determining how the amount and quality of stored C is affected by management strategies intended to increase SOM content.

SOM quality has been found to be relevant in determining organic C persistence and turnover, but recent experimental advances have demonstrated that molecular structure alone does not control SOM stability, since environmental and biological controls predominate (Schmidt et al., 2011). In fact, the chemical recalcitrance of soil C compounds to microbial attack is an intrinsic property affecting SOM stability, on the other hand soil organic matter turnover can be mainly governed by microbial accessibility not recalcitrance (Dungait et al., 2012). Management practices can have significant implications for soil quality and C sequestration potential in agricultural soils. Among the RMPs (reduced management practices), reduced or zero tillage, application of biosolids, cover and deep-rooting crops, conversion to grasslands and woodland, rotations, fertilization and irrigation are the most effective (Campiglia and Cavalieri, 2008). The same authors conclude that among the diverse cropping systems, conservative agriculture rather than conventional or organic agriculture provides a better storage and utilization of C.

Land use, plant cover and agricultural systems (Moscatelli et al., 2007; Marinari et al., 2010a and b; Lagomarsino et al., 2008) determine the quality of organic C reaching the soil and forming organic matter. Plant debris and litter derived from conifer trees, for example, are characterized by recalcitrant compounds such as lignins, waxes, resins that will constitute very stable SOM pools (Moscatelli et al., 2007). Marinari et al., (2007) reported significant changes in SOM quality using pyrolisis technique in a study comparing organic and conventional farming systems. The increase of soil microbial biomass and activity strictly was related to pyrolytic mineralization and humification indices of SOM in organic soils. Mineral fertilization caused a greater alteration of native SOM with a release of water soluble organic C and relatively large amounts of aliphatic pyrolytic products.

Other factors linked to land use and management may influence SOM stability and properties: organic C interactions with mineral constituents may in fact prevent its accessibility and consequent decomposition (Sollins et al., 1996).

The recent literature has focused on physical and/or chemical fractionation of soil that may distinguish SOM pools stability, C residence time and, at the same time, those fractions functionally meaningful in soil biogeochemical cycling (Six et al., 2002). Micro-scale investigations allow a deeper understanding of processes underlying the relationships between organic matter, minerals and microorganisms in soil, and help to understand modifications induced by soil management. SOM is heterogeneously distributed among different-sized primary particles and can respond differently to soil disturbance and management changes. Biologically active or labile C fractions of SOM are likely controlled by management to a much greater extent than total SOM. Slower C turnover rates in the clay than in the sand fraction were explained by SOM quality change, increased spatial inaccessibility and the adsorption of organic compounds on mineral surfaces (von Lützow et al., 2007).

Finally soil structure, as aggregation of particles, is an important soil property, mediated by SOM. In particular, SOM enhances aggregation through the bonding of primary soil particles; thus any management practices that usually increases SOM will ultimately improve soil structure and stability (Bronick and Lal, 2005).

Nutrient Cycling

The correct management of nutrient supply to crops is certainly one of the most important priorities in securing world food supply without causing unacceptable environmental impacts (Powlson et al., 2011). In relation to different environmental conditions and crop systems worldwide, we range from chronic deficiencies (unsustainable agricultural systems due to great removal of nutrients) to vast over supplies with consequent waste of resources and environmental pollution (Powlson et al., 2011). Major issues are related to nitrogen (N), being N cycle extremely complex and consisting in several forms that can be lost from the soil (leached as nitrate or lost as gases such as N_2, N_2O, NH_3, NO_2 etc.) with significant ecological impacts such as: nitrate contamination of water, eutrophication of lakes and rivers, soil acidification. In particular nitrogen losses as N_2O are of concern representing a significant contribution to overall greenhouse gases and having a global warming potential much higher than that of CO_2. Powlson et al., (2011) claim for innovative agronomic management optimizing recycling practices, improvement of N fixation, intercropping/agroforestry and mainly a better exploitation of soil potential to provide inorganic N.

N distribution in soil fractions, and then its fate, is deeply affected by multiple factors such as soil physico-chemical properties (pH and redox conditions being the most important) and microbial processes, in particular those performed by nitrogen fixing microrganisms and ammonia oxidizers. N mineralization processes, from native organic matter or from organic fertilizers, is an important indicator of soil quality aimed to assess soils capacity to supply inorganic N to crops. Marinari et al. (2010b), in a three-year crop rotation, found that monitoring N mineralization kinetics proved to be effective in discriminating between organic and conventional cropping systems in terms of mineralization–immobilization processes during the crop rotation period. Moreover, it offered potential to assess the most suitable N release patterns of fertilizers used under organic cropping system.

In forest ecosystems, fertilization of forest plantations has become increasingly important as an intensive management tool. Responses to fertilization are generally positive as far as above ground biomass is concerned whereas more uncertain is the response of belowground pools and processes (Moscatelli et al., 2008). In the same study, performed in a short rotation poplar plantation, short- and medium-term effects of fertilization on soil C and N pools and cycling were reported with a consistent immobilization process of ammonium and increase of nitrate.

A meta-analysis based on 107 datasets from 64 long-term trials from around the world revealed that, over the long term, mineral fertilizer application increased soil organic carbon content and consequently microbial C by 15.1% with respect to unfertilized control treatments (Giesseler and Scow, 2014). On the other hand, long-term repeated mineral N applications can alter microbial community composition being these changes mainly linked to variations of soil pH.

SOIL BIOLOGICAL PROPERTIES – SOIL BIOTA – BIODIVERSITY

Soil system is distinct from any other inert material because of its dynamic nature which is certainly the most interesting feature of this environmental resource. Soil biota is the main responsible of changes, adaptations, transformations, evolution and, in general, of almost all processes taking place within soil environment. The soil fauna and microbial community are considered critical in any ecosystem, by acting on the decomposition of soil organic matter, nutrient cycling, and affecting the soil chemical and physical properties, with ultimately determine soil fertility and sustainability. The biological component of the soil, due to its high turnover rate, is rapidly affected by soil use and management or any other disturbance, differently from most chemical and physical properties, which take longer times to be changed (Pankhurst et al., 1995). Soil microbial biomass is, in fact, considered an early warning of degration process and for this reason is definitely one of the best dynamic indicators of soil quality. Edaphic organisms are therefore good indicators of soil quality and health, especially if the indicator corresponds to specific ecological processes occurring in soil (Cardoso et al., 2013).

Soil microbial biomass is also characterized by genetic, structural or functional diversity whose level can be deeply modified, most of the time lowered, by management, however leaving unaltered biomass size values (Pignataro et al., 2012). Belowground biodiversity represents one of the largest reservoirs of biological diversity on earth, thus how biodiversity loss impacts ecosystem functioning is a priority of current research (Bardgett and Van der Putten, 2014).

In recent years, the attention has therefore been shifted on changes in soil biodiversity more than in populations size in response to soil management. Maintaining a high microbial diversity has got important ecological implication representing a chance to keep essential ecosystem processes after a disturbance (Kennedy, 1999). Such capacity is also defined as resilience which means a biological buffering against disturbances or stress in an ecosystem. The resilience capacity of soil microbial biomass has been tested in response to heat stress in soils sampled from long term agriculture and permanent grassland (Riah-Anglet et al., 2015). The amplitude of the negative or positive responses to heat of enzyme activities and microbial biomass was smaller in grassland soils pointing to a more resistant system to drastic stress due to its highly abundant and highly diversified microbial community. The soil management history explained thus the different resilience capacity of these microbial groups to heat stress as the changes in the microbial community and functions were more pronounced in the conventionally cropped soils.

Human agricultural activity has incredible impacts on soils and soil biota. Among soil management practices, tillage, agrochemicals, organic amendments etc. may deeply affect soil biological properties. Investigating a gradient of agricultural land management, Culman et al., (2010) found that agricultural intensification leads to losses in aboveground and belowground biodiversity. Species biodiversity is affected by management practices; generally high-input agricultural practices decrease biodiversity while the low-input practices enhance biodiversity (Munyanziza et al., 1997; Lupwayi et al., 2001).

From the microbial structural point of view, agricultural practices generally lead to a shift from fungal-based soil food webs to more bacterial-based soil food webs. Especially under tillage, fungi, including AMF (arbuscular mycorrhizal fungi), with their extensive hyphal network are reduced. Vice-versa, fungal biomass increases within years after tillage abandonment (Thiele Brun et al., 2012).

Tillage modifies the physical and chemical environment of the soil and affects soil water content, soil temperature, aeration and the contact between SOM and mineral particles, globally influencing the soil microbial population which in turn affects the physico-chemical environment of the soil (Marinari et al., 2015). The adoption of no-tillage (NT) in cropping systems in China over 1.33 million hectars showed numerous beneficial impacts on soil quality in terms of physical, chemical and, mainly, biological properties (Huang et al., 2013). Positive effects of NT have been reported on microbial diversity, counts and activity although the authors claim for further specific investigation in rice cropping systems. However it should also be considered that for many pests tillage may have both beneficial and detrimental effects; on this purpose Roger-Estrade et al., (2010) claim for a global consideration of soil management and experimental design taking into account all aspects of soil ecology.

The advantages of no tillage (NT) for improving soil quality and microbial biomass are greatly enhanced by growing cover crops and mulches. In the Mediterranean environment dead mulches are generally cut during spring, just before cultivating the main crop and are left on the soil until the end of the growing season, thus causing a reduction in soil water evaporation, a decrease in daily soil temperature excursion and weed control. Soil microorganisms benefit of these improved soil physical conditions; a positive effect of mulching on soil microbial biomass in tomato crop in central Italy was indeed reported by Marinari et al., (2015).

Extensive literature showed that organic amendments such as municipal solid wastes, sewage sludge, compost, and manure enhance soil microbial biomass C, basal respiration, and enzyme activities, with variations in soil microbial community structure (Bastida et al., 2008; Ros et al., 2008). Organic fertilization and amendments are the fundamentals of organic agriculture. Several studies supported the evidence of an increase of microbial biomass and activity under organic management, leading to high nutrients availability for plants (Tu et al., 2003; Wang et al., 2004). Tu et al., (2003) showed that enhanced soil microbial biomass and activity were associated with high net N mineralization rates. In separate studies, Monokrousos et al., (2006) and Melero et al., (2006) found the activities of enzymes involved in the cycles of N and P (phosphorus) enhanced through the organic management, with beneficial effects for soil nutrients supply. In particular, the addition of substrates of differentiated chemical structure may foster an increase of microbial diversity in organically managed soil because a diverse microbial community is able to better transform C from organic debris into biomass.

In forest ecosystems, soil microbial community may respond in different ways to anthropic disturbances such as prescribed burning, erosion, harvesting, heavy traffic, changes in plant diversity (Cambi et al., 2015; Giai and Boerner, 2007; Lalor et al., 2007; Leckie, 2005). The clearing of trees, in particular, causing the restructuring of vegetation, modification in litter quality and quantity, alteration of root exudates, leaching of plant nutrients and changes in the microclimate, can create conditions capable of affecting soil biota and influencing microbial communities diversity (Marshall, 2000).

It is clear that when primary tropical forest is converted to agriculture, many groups of organisms (especially macro- and micro-fauna) are dramatically reduced in diversity in addition to population size (Wall et al., 2010). Pignataro et al., (2012) reported, after an oak forest coppicing, a significant increase of soil microbial pool, respiration and changes in diversity patterns promoted by plant debris and rhizodepositions released after cutting. Significant positive effects on microbial biomass C and N are reported by Wen et al., (2014) showing beneficial impacts when turning from monocultures into mixed forests as a forest management practice.

Soil Structure – Aggregation – Erosion

Erosion rates are strictly linked to soil structure stability and this depends on land use and cover, cropping systems leaving bare soil, intensive tillage, landslides, wildfires, etc. Virto et al., (2015) provided, on this purpose, an exhaustive report, specifically on Western Europe soils, on the causes of physical degradation, suitable soil quality indicators to be used and all possible actions to undertake within the EU.

Soil structure can be defined as the aggregation, rearrangement, flocculation and cementation of particles mediated by SOM, biota, ionic bridging, clay and carbonates

(Bronick and Lal, 2005). Changes in soil structure and aggregation level can strongly modify SOM properties and C residence time in soil. In particular the physical disconnection between decomposers and organic matter, deriving from modification of soil structure, is likely to be one reason for persistence of SOM (Schmidt et al., 2011).

Tillage systems based on plowing and disking (so-called conventional tillage) in the tropical regions decrease soil organic matter and increase the erosion process, leading to chemical, physical, and biological changes in the soil characteristics that enhance the dependence on external inputs and consequently enhance production costs, causing environmental impacts. On the other hand, less impacting cropping systems like no-tillage and organic farming are much more reliant on biological processes for sustainability (Cardoso et al., 2013).

The decline in soil structure is increasingly seen as a form of soil degradation (Chan et al., 2003) and is often related to land use and soil/crop management practices. Soil structure influences soil water movement and retention, erosion, crusting, nutrient recycling, root penetration and crop yield. Moreover, under no-tillage management, the amount of water-stable macroaggregates is generally greater and the rate of macroaggregate turnover is approximately half the rate of that under conventional tillage (Six et al., 1999, 2000). Accordingly, greater mean residence times and accumulation of soil C in the surface layer of no-tilled soil are expected (Six et al., 2002). Therefore, soil structure can be significantly modified through management practices and environmental changes. Practices that usually increase productivity and decrease soil disruption enhance aggregation and structural development (Bronick and Lal, 2005). In this regard, improved management practices include tillage methods, residue management, amendments, soil fertility management and nutrient cycling.

In a grassland ecosystem Cui and Holden (2015) reported significant interactions among soil structure, different types of management practices (reseeding, fertilization, stocking rate) and organic matter. The data obtained, through an on field *in situ* assessment, indicated a strong correlation between structure and enzyme activity, both of which were influenced by the diverse management practices. SOM contributed to this correlation, acting as a two-way 'bridge' between structure and enzyme activity.

REFERENCES

Bardgett R. D. and van der Putten W. H., (2014). Belowground biodiversity and ecosystem functioning. *Nature*, 515, 505–511.

Bastida F., Zsolnay A., Hernández T., García C., (2008). Past, present and future of soil quality indices: a biological perspective, *Geoderma*, 147, 159-171.

Bronick, C. J., Lal, R., (2005). Soil structure and management: a review. *Geoderma* 124 (1–2), 3–22.

Cambi M., Certini G. Neri F., Marchi E., (2015). The impact of heavy traffic on forest soils: A review, *Forest Ecology and Management* 338, 124–138.

Campiglia E. and Cavalieri A., (2008). The role of organic practices on soil organic carbon sequestration in arable land in "Soil carbon sequestration under organic farming in the

Mediterranean environment" (S. Marinari and F. Caporali eds), Transworld Research Network, Kerala, India. ISBN 978-81-7895-327-4.

Canadell J. G., Mooney H. A., Baldocchi D. D., Berry J.A., Ehleringer J. R., Field C. B., Gower S. T., Hollinger D. Y., Hunt J. E., Jackson R. B., Running S.W., Shaver G. R., Steffen W., Trumbore S. E., Valentini R., Bond B. Y., (2000). Carbon metabolism of the terrestrial biosphere: a multi-technique approach for improved understanding. *Ecosystems* 3, 115–130.

Cardoso E. J. B. N., Figueiredo Vasconcellos R.L., Bini D., Yumi Horta Miyauchi M., Alcantara dos Santos C., Lopes Alves P.R., Monteiro de Paula A., Shigueyoshi Nakatani A., de Moraes Pereira J., Nogueira M.A., (2013). Soil health: looking for suitable indicators. What should be considered to assess the effects of use and management on soil health? *Sci. Agric.* 70, 4, 274-289.

Chan, K. Y., Heenan, D. P., So, H.B., (2003). Sequestration of carbon and changes in soil quality under conservation tillage on light textured soils in Australia: a review. *Aust. J. Exp. Agric.* 43, 325–334.

Cui J. and Holden N.M., (2015). The relationship between soil microbial activity and microbial biomass soil structure and grassland management. *Soil & Tillage Research* 146 32–38.

Culman S. W., Young-Mathews A., Hollander A.D., Ferris H., Sanchez-Moreno S., O'Geen A. T., Jackson L. E. (2010). Biodiversity is associated with indicators of soil ecosystem functions over a landscape gradient of agricultural intensification. *Landscape Ecol*, 25:1333-1348.

Doran, J. W., Parkin, T. B., (1994). Defining and assessing soil quality. In: Doran, J. W., Coleman, D.C., Bezdicek, D.F., Stewart, B.A. (Eds.), Defining soil quality for a sustainable environment. SSSA, Spec Pub no 35, Madison, pp 3–21.

Dungait, J. A. J., Hopkins, D. W., Gregory, A. S., Whitmore, A. P., (2012). Soil organic matter turnover is governed by accessibility not recalcitrance. *Global Change Biology* 18 (6), 1781–1796.

Geisseler D., Scow K. M., (2014). Long-term effects of mineral fertilizers on soil microorganisms: A review, *Soil Biology & Biochemistry* 75, 54-63.Gil-Sotres, F., Trasar-Cepeda C., Leiros M.C., Seoane S. (2005). Different approaches to evaluating soil quality using biochemical properties. *Soil Biol Biochem*, 37, 877–887.

Giai, C., Boerner, R. E. J., (2007). Effects of ecological restoration on microbial activity, microbial functional diversity, and soil organic matter in mixed-oak forests of southern Ohio, USA. *Appl. Soil Ecol.* 35, 281–290.

Guo, L. B., and Gifford, R. M., (2002). Soil carbon stocks and land use change: a meta analysis. *Global Change Biology*, 8: 345-360.

Huang Min, Ligeng Jiang, Yingbin Zou, Shihong Xu and Guofu Deng (2013). Changes in soil microbial properties with no-tillage in Chinese cropping systems. *Biol Fertil Soils* 49:373–377, DOI 10.1007/s00374-013-0778-6.

Kennedy A.C., (1999). Bacterial diversity in agroecosystems, *Agriculture, Ecosystems & Environment* 74, 1–3, 65–76.

Lagomarsino, A., Di Tizio, A., Marinari, S., Moscatelli, M.C., Mancinelli, R., Grego, S. (2008). Soil organic matter pools under different system management and tillage level in a three-year crop rotation. *Agrochimica* 52 (6), 395-406.

Lalor, B. M., Cookson, W.R., Murphy, D. V., (2007). Comparison of two methods that assess soil community level physiological profiles in a forest ecosystem. *Soil Biol. Biochem.* 39, 454–462.

Leckie, S.E., (2005). Methods of microbial community profiling and their application to forest soils. *For. Ecol. Manag*. 220, 88–106.

Lupwayi, N. Z., Arshad, M.A., Rice, W. A., Clayton, G. W., (2001). Bacterial diversity in water-stable aggregates of soils under conventional and zero tillage management. *Appl. Soil Ecol.* 16, 251– 261.

Marinari S., Lagomarsino A., Moscatelli M.C., Di Tizio A., Campiglia E., (2010a). Soil carbon and nitrogen mineralization kinetics in organic and conventional three-year cropping systems, *Soil & Tillage Research* 109, 161–168.

Marinari S., Mancinelli R., Brunetti P., Campiglia E., (2015). Soil quality, microbial functions and tomato yield under cover crop mulching in the Mediterranean environment, *Soil & Tillage Research* 145, 20–28.

Marinari, S., Liburdi, K., Fliessbach, A., Kalbitz, K., (2010b). Effects of organic management on water-extractable organic matter and C mineralization in European arable soils. *Soil and Tillage Research* 106 (2), 211-217.

Marinari, S., Masciandaro, G., Ceccanti, B., Grego, S., (2007). Evolution of soil organic matter changes using pyrolysis and metabolic indices: a comparison between organic and mineral fertilization. *Bioresource Technology* 98 (13), 2495-2502.

Marshall, V. G., (2000). Impacts of forest harvesting on biological processes in northern forest soils. *For. Ecol. Manag*. 133, 43-60.

Melero, S., Ruiz Porras, J. C., Herencia, J. F., Madejon, E., (2006). Chemical and biochemical properties in a silty loam soil under conventional and organic management. *Soil and Tillage Research* 90, 162-170.

Monokrousos N., Papatheodorou E. M., Diamantopoulos J.D., Stamou G.P., (2006). Soil quality variables in organically and conventionally cultivated field sites, *Soil Biology & Biochemistry* 38, 1282–1289.

Moscatelli M.C., Di Tizio A., Marinari S., Grego S., (2007). Microbial indicators related to soil carbon in Mediterranean land use systems, *Soil Till Res*, 97, 51–59.

Moscatelli M.C., Lagomarsino A., De Angelis P., Grego S., (2008). Short- and medium-term contrasting effects of nitrogen fertilization on C and N cycling in a poplar plantation soil, *Forest Ecology and Management,* 255, 447–454.

Munyanziza, E., Kehri, H. K., Bagyaraj, D. J., (1997). Agricultural intensification, soil biodiversity and agro-ecosystem function in the tropics: the role of mycorrhiza in crops and trees. *Appl. Soil Ecol.* 6, 77– 85.

Pankhurst, C. E., Hawke, B. G., McDonald, H. J., Kirkby, C. A., Buckerfield, J. C., Michelsen, P., O'Brien, K. A., Gupta, V. V. S. R., and Doube, B. M. (1995). Evaluation of soil biological properties as potential bioindicators of soil health. *Australian Journal of Experimental Agriculture* 35:1015-1028.

Pereira e Silva M.C., Semenov A.V., Schmitt H., van Elsas J. D., Falcão Salles J., (2012). Microbe-mediated processes as indicators to establish the normal operating range of soil functioning, *Soil Biology and Biochemistry*, 1-2.

Pignataro A., Moscatelli M.C., Mocali S., Grego S., Benedetti A., (2012). Assessment of soil microbial functional diversity in a coppiced forest system, *Applied Soil Ecology*, 62, 115-123.

Powlson D.S., Gregory P.J., Whalley W.R., Quinton J.N., Hopkins D.W., Whitmore A.P., Hirsch P.R., Goulding K.W.T., (2011). Soil management in relation to sustainable agriculture and ecosystem services, *Food Policy,* 36, S72–S87.

Riah-Anglet W., Trinsoutrot-Gattin I., Martin-Laurent F., Laroche-Ajzenberg E., Norini M. P., Latour X., Laval K., (2015). Soil microbial community structure and function relationships: A heat stress experiment. *Appl Soil Eco,* 86, 121–130.

Roger-Estrade J, Anger C, Bertrand M, Richard G., (2010). Tillage and soil ecology: partners for sustainable agriculture. *Soil Till Res*, 111:33-40.

Ros, M., Goberna, M., Pascual, J.A., Klammer, S., Insam, H., (2008). 16S rDNA analysis reveals low microbial diversity in community level physiological profile assays. *J. Microbiol. Methods.* 72, 221-226.

Schmidt M.W. I., Torn M.S., Abiven S., Dittmar T., Guggenberger G., Janssens I.A, Kleber M., Kögel-Knabner I., Lehmann J., Manning D.A.C., Nannipieri P., Rasse D.P., Weiner S.,Trumbore S.E. (2011). Persistence of soil organic matter as an ecosystem property. *Nature* 478, 49–56.

Six, J., Conant, R.T., Paul, E.A. & Paustian, K., (2002). Stabilization mechanisms of soil organic matter: implications for C-saturation of soils. Plant and Soil, 241, 155–176.

Six, J., Elliott, E. T. & Paustian, K., (2000). Soil macroaggregate turnover and microaggregate formation: a mechanism for C sequestration under no-tillage agriculture. Soil Biology and Biochemistry, 32, 2099–2103.

Six, J., Elliott, E.T. & Paustian, K., (1999). Aggregate and soil organic matter dynamics under conventional and no-tillage systems. Soil Science Society of America Journal, 63, 1350–1358.

Sollins, P., Homann, P. & Caldwell, B. A., (1996). Stabilization and destabilization of soil organic matter: mechanisms and controls. *Geoderma* 74, 65–105 (1996).

Thiele-Bruhn S., Bloem J., de Vries F. T., Kalbitz K. and Wagg C., (2012). Linking soil biodiversity and agricultural soil management, *Current Opinion in Environmental Sustainability*, 4:523–528.

Tu, C., Koenning, S. R., Hu, S., (2003). Root-parasitic nematodes enhance soil microbial activities and nitrogen mineralization. *Microb. Ecol.* 46, 134–144.

Virto I., Imaz M. J., Fernández-Ugalde O., Gartzia-Bengoetxea N., Enrique A., Bescansa P., (2015). Soil Degradation and Soil Quality in Western Europe: Current Situation and Future Perspectives, *Sustainability*, 7, 313-365; doi:10.3390/su7010313

von Lützow, M., Kögel-Knabner, I., Ekschmitt, K., Flessa, H., Guggenberger, G., Matzner, E., Marschner, B., (2007). SOM fractionation methods: Relevance to functional pools and to stabilization mechanisms. *Soil Biology and Biochemistry* 39(9), 2183–2207.

Wall, D.H., Bardgett, R.D., Kelly, E.F., (2010). Biodiversity in the dark. *Nature Geoscience* 3, 297–298.

Wang, W. J., Smith, C. J., Chen, D., (2004). Predicting soil nitrogen mineralization dynamics with a modified double exponential model. *Soil Sci. Soc. Am. J.* 68, 1256.

Wen L., Lei P., Xiang W., Yan W., Liu S., (2014). Soil microbial biomass carbon and nitrogen in pure and mixed stands of Pinus massoniana and Cinnamomum camphora differing in stand age, *Forest Ecology and Management* 328, 150–158.

In: Soil Management
Editor: Manuel Esteban Lucas Borja
ISBN: 978-1-63483-274-8

Chapter 2

BIODIESEL AND DIESEL BLENDS PHYTOTOXICITY IN SOIL BIODEGRADATION

Ederio Dino Bidoia, Ivo Shodji Tamada, Paulo Renato Matos Lopes, Jaqueline Matos Cruz and Renato Nallin Montagnolli
UNESP, São Paulo State University, Bioscience Institute Rio Claro, São Paulo, Brazil

ABSTRACT

Oil spills cause substantial damage to the environment. Accidents in the petroleum industry create huge public concern, requiring effective responses from governments and corporations. There are also chronic impacts involving petroleum hydrocarbons, such as leaks in storage tanks, although they do not stand out as much in public opinion. Thus, continuous and prolonged pollution provides high environmental persistence of these compounds in soil, surface, and groundwater. The massive energy consumption based on non-renewable fossil fuels has led to the search for new sustainable and less polluting sources. Therefore, a viable alternative diesel fuel is biodiesel, originated from renewable sources, either plant or animal. Conventional physical soil treatment comprises of contaminants separation mostly, without any chemical degradation or transformation. These processes have many limitations and high cost. Most hydrocarbons are usually absorbed within the soil matrix by lowering its efficiency removal and absorption. Biological processes, in turn, are promising decontamination technologies, especially by combining simplicity and cost-effectiveness. Hence, bioremediation emerges as the least aggressive and best-suited eco-friendly technology, among the alternatives. Due to increasing use of biodiesel and bioremediation as a treatment tool, this study aimed to evaluate the phyto-toxicity in soil contaminated by diesel, biodiesel, and mixtures during different times of degradation. Seeds of *Eruca sativa* (arugula) and *Lactuca sativa* (lettuce) were used to observe inhibition of germination and seedling development, evaluated by germination index. Toxicity bioassays using biodiesel as a contaminant of animal or vegetable origin showed increased toxicity in all organisms tested over time of biodegradation in soil. The study showed that although biodiesel is a renewable energy source with great ecological appeal, some major precaution must be taken. Even with favorable results for their higher biodegradability, other parameters must be considered to make biodiesel safe and ecologically viable by monitoring their toxicity.

Keywords: ecotoxicology, hidrocarbon, bioremediation, germination

INTRODUCTION

According to the International Energy Agency (IEA, 2011), natural gas, oil and their derivatives are the largest currently used source of energy. These sources, however, are not renewable. The massive energy consumption based on non-renewable fossil fuels has led to the search for new less polluting sustainable sources. Due to the scarcity of these fossil fuels, biodiesel arises as a viable alternative source of energy (Pinto et al., 2005; Geris et al., 2007). Biodiesel is relevant as it enters economic systems as a potential substituent capable of meeting high energy demands (Geris et al., 2007).

The Brazilian National Council for Energy Policy (CNPE) established by Resolution 6 in September 16, 2009 a regulation that from January 2010 onwards all diesel sold in Brazil must present 5% biodiesel in its composition. Though the new policy raised awareness that renewable resources are top priority, another problem arises due to the great production and consumption of biodiesel, which may lead to hazardous effects: environmental contamination.

Oil spills cause substantial damage to the environment. Accidents in the petroleum industry create huge public concern, requiring effective responses from governments and corporations. There are also chronic impacts involving petroleum hydrocarbons, such as leaks in storage tanks, although they do not stand out as much in public opinion. Thus, continuous and prolonged pollution provides high environmental persistence of these compounds in soil, surface, and groundwater.

Any substance can be a toxic agent. Depending on the exposure conditions, adsorbed dose, frequency, exposure time and in which way biodiesel is applied, it can also be a toxic agent (Barros and Davino, 2003). The environmental toxicology field, which is very close to ecotoxicology in its definition, studies the harmful effects of chemical agents, such as biodiesel, in living organisms taking into account ecological mechanisms (Passarelli, 2003).

Phytotoxicity assays are the most commonly performed tests by evaluating the germination and seed growth of different plant species exposed to pollutants. The germination and seed growth is held in soil or hydroponic systems, but can also be adapted to other media and substances (Suter, 1993).

Ecotoxicology

Toxicology is a major science field that investigates adverse effects of chemicals in the organisms due to diverse types of harmful interactions (Zagatto and Bertoletti, 2006; Oga and Siqueira, 2003), varying according to the type of exposure. Research on this topic goes back to the course of the human history since ancient times, even though it did not receive such terminology at that time. It was present in the human civilization through the knowledge of the poisonous effects of some animals and varieties of plants in human health. Ancient documents show that use of toxic substances since around 1500 BC, with a list of 800 active ingredients, including metals, animal poisons and many toxic plants (Zagatto and Bertoletti, 2006; Oga and Siqueira, 2003).

Toxicology approaches are connected to how the contaminants affect their environment. Investigation can either focus on the nature of the toxic agent or the way it harms a biological system. Recently, most relevant studies arise from areas associated with environmental, occupational, food, drugs and cosmetics and social toxicology (Oga and Siqueira, 2003).

Within toxicology studies, there are two very similar fields of study: environmental toxicology, which studies the toxic effect of an agent for a species or a group of living organisms; and ecotoxicology, which studies the toxic agent in a more integrated way when released into different types of environment, taking into account the entire ecosystem (Passarelli, 2003). According to Zagatto and Bertoletti (2006), the concept of ecotoxicology was much discussed at the end of the 70's and currently fundamentally described the interaction of natural or synthetic substances on living organisms, populations, communities, animal or vegetable, terrestrial or aquatic(Landis and Yu, 1998). Together they constitute the biosphere as a base model for interactions with the environment.

Biodiesel

With the increasing scarcity of fossil fuels, there is no consensus for one solution to avoid an energy crisis. However, the search for renewable resources, such as biodiesel, is widely debated. Therefore, the Brazilian National Council for Energy Policy (CNPE) gradually introduced increasingly amounts of biodiesel in diesel sold in Brazil. New government resolutions matched the mixture percentages found in European countries, which is around 5% (Pinto et al., 2005).

Biodiesel is subject to abiotic degradation, including photo-oxidation, which involves addition of oxygen to the unsaturated fatty acids (Ferrari and Souza, 2009). As an organic compound, biodiesel is also subject to biodegradation. The overall formulation may change along this process, as well as their environmental behavior.

Thus, this chapter determined the synergistic effects of biotic and abiotic degradation over time from exposure to soil towards the ecotoxicity of biodiesel and diesel to plants seeds and seedlings. The assessment of phytotoxicity quantified growth inhibition of arugula (*Eruca sativa*) and lettuce (*Lactuca sativa*).

METHODS

Soil and Contaminant Samples

The base soil samples were collected in the experimental garden at the Biosciences Institute of Universidade Estadual Paulista (UNESP), Rio Claro campus (22°24'39"S, 47°33'39"W). Soil microorganisms used in all analyzes were indigenous to the collected soil, obtained in the experimental garden. The area had a history of periodic petroleum products release. The microorganisms were assumed to contribute to the biodegradation process of hydrocarbon contaminants along time.

Table 1. Toxicity bioassays identification

Assay Identification	Content
D	Diesel
BA	Biodiesel from animal source
BV	Biodiesel from vegetable source
CN	Negative control with water
CP	Positive control with ZnSO4

All the treatments in the phytotoxicity assays followed the nomenclature given in Table 1. Biodiesel from animal and vegetable source used in this study were provided by JBS and Caramuru Foods, respectively. The diesel oil used in this work was donated by Replan/Petrobras. Regarding the animal biodiesel, it is a blend composed of approximately 33% of vegetable biodiesel (soy oil) and 67% biodiesel from animal (beef tallow). However, for nomenclature purposes, it was referred as BA (biodiesel of animal origin).

Biodegradation of Contaminants in Soil

For the preparation of a soil biodegradation environment, it was necessary to conduct a pre-selection of microorganisms capable of biodegrading the desired contaminants (biodiesel and diesel). We made a crescent gradient of diesel/biodiesel blends to promote a biodegrading environment base inoculum as follows: in a plastic bag with 3 kg of previously sieved sand (sieve with a 1.0 mm granulometry) measured by semi-analytical scale (Shimadzu BL3200H), 50 ml of diesel, biodiesel or biodiesel / diesel blend (Table 1) and 100 ml of distilled water. The bag was closed and homogenized. After this stage, the bag was perforated with small holes approximately 1 mm and spacing of about 1 cm each and buried 15 cm deep 30 days (Lopes et al., 2010; Lopes and Bidoia, 2009; Cetesb, 1990).

Phytotoxicity Assays

Phytotoxicity assays were conducted to observe the variation of toxicity during the biodegradation of contaminants in three different soil samples during the degradation time course. Diesel and biodiesel from vegetable and animal source were analysed separately. The toxicity experiments were based on methodologies proposed by Lopes et al. (2010), Morales, (2004), Tiquia et al., (1996), Wang (1987), and Tamada et al., (2012).

For the phytotoxicity bioassays we used *Eruca sativa* (arugula) and *Lactuca sativa* (lettuce) seeds. All seeds used in bioassays were from the same batch (#020020). The batch had a stock germination ratio of 85% and a purity of 99%.

Toxicity bioassays were periodically every 30 days of the biodegradation in soil. Six analyzes were performed toxicity according to the time of biodegradation of the contaminants in soil: initial time (TO), 30 days (T30), 60 days (T60), 90 days (T90), 120 days (T120) and 150 days (T150). Bioassays were mounted with mixtures containing sieved sand (1.0 mm granulometry) with blends of contaminants (D, BA, BV) and distilled water in the ratio 100 g:10 ml:5 ml, respectively. This mixture was buried and aliquots were taken every 30 days for the bioassays. For each treatment were used plastic cups (50 ml cup) in triplicates, containing 50.00 g of the soil. Each replica was made with 10 seeds of *E. sativa* or *L. sativa* and, 2 ml of deionized water by covering the soil with a PVC film. The bioassays were held for 120 h in a growth chamber (BOD) by Solab SL224 at 22°C and thereafter had the germination quantified. It was also simultaneously mounted a positive control with addition of 0.05mL of 2.0 M zinc sulfate and a negative control containing only the ground, without any contaminant to test the viability of the seeds used in all the phytotoxicity bioassay.

Data Analysis

For the descriptive analysis of bioassays the following parameters were considered toxicity indicators: hypocotyl and root elongation and the germination index (GI, ratio of germination and root growth compared to the negative control), as shown in Equation 09 (Morales, 2004; Tiquia et al., 1996). This methodology is a practical resource, based on low cost and reasonable sensitivity in quantification of toxic biological inhibitors (Inazaki et al., 2001).

$$GI = \frac{\left(\frac{Gt}{Gc}\right) \times \left(\frac{Et}{Ec}\right)}{100} \qquad \text{(Equation 1)}$$

where GI = germination index, Gt = germination in a certain treatment, Gc = germination in the control assay, Et = elongation of vegetable tissue (root and hypocotyl) in a certain treatment, Ec = elongation of vegetable tissue in the control assay.

METHODS

The phytotoxicity results of the bioassays are shown in Figures 1 and 2. The data analysis is based on the inhibition of root and hypocotyl growth compared to the negative control and the germination index (GI).

Eruca sativa Bioassays

The results presented phytotoxicity response in bioassay using *E. sativa* seeds (Figure 1). According to the calculated germination index (GI), moderate toxicity levels occurred in all starting time points (T0), and decreased after 30 days. At time T60 the diesel (D) GI presented low toxicity, with a GI greater than 100% (Figure 1). Low toxicity was also found in biodiesel of animal origin (BA), but with less intensity (67.43% GI).

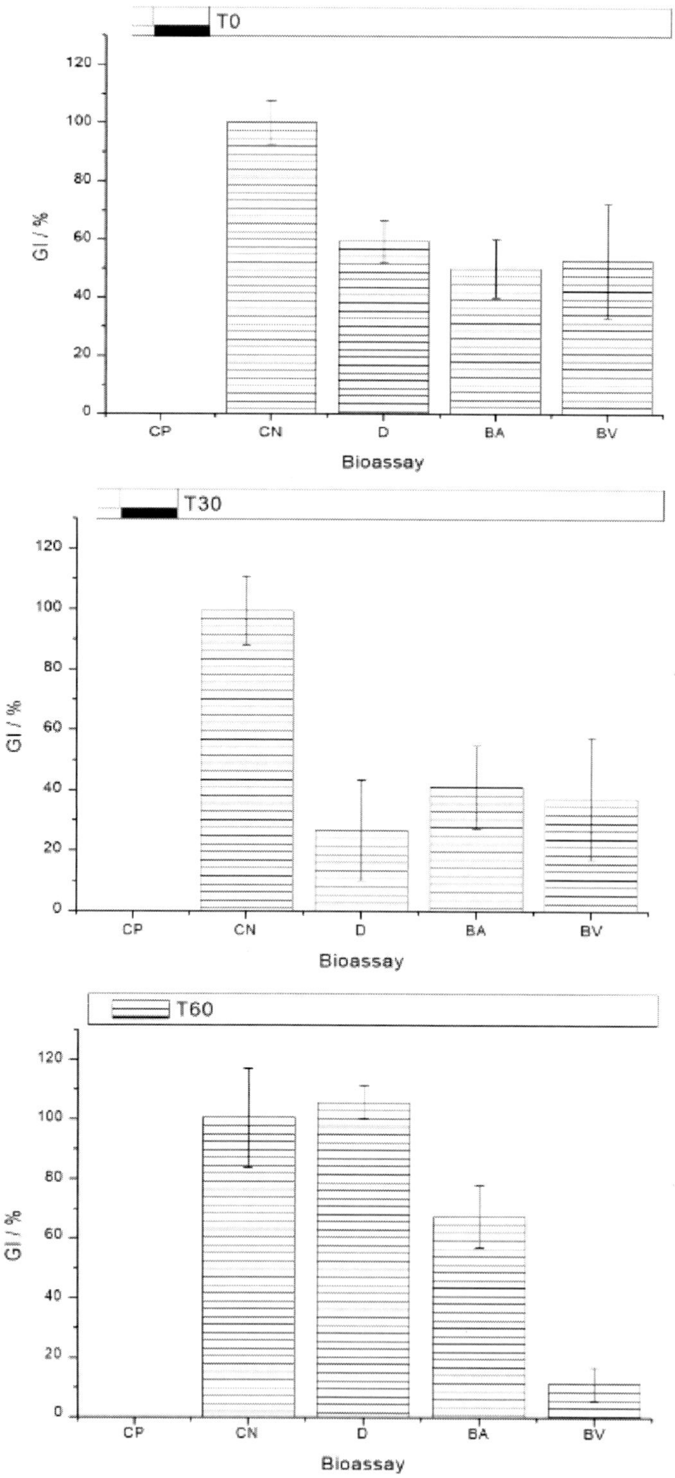
T0
GI / %
0
20
40
60
80
100
120
CP
CN
D
BA
BV
Bioassay
T30
GI / %
0
20
40
60
80
100
120
CP
CN
D
BA
BV
Bioassay
T60
GI / %
0
20
40
60
80
100
120
CP
CN
D
BA
BV
Bioassay

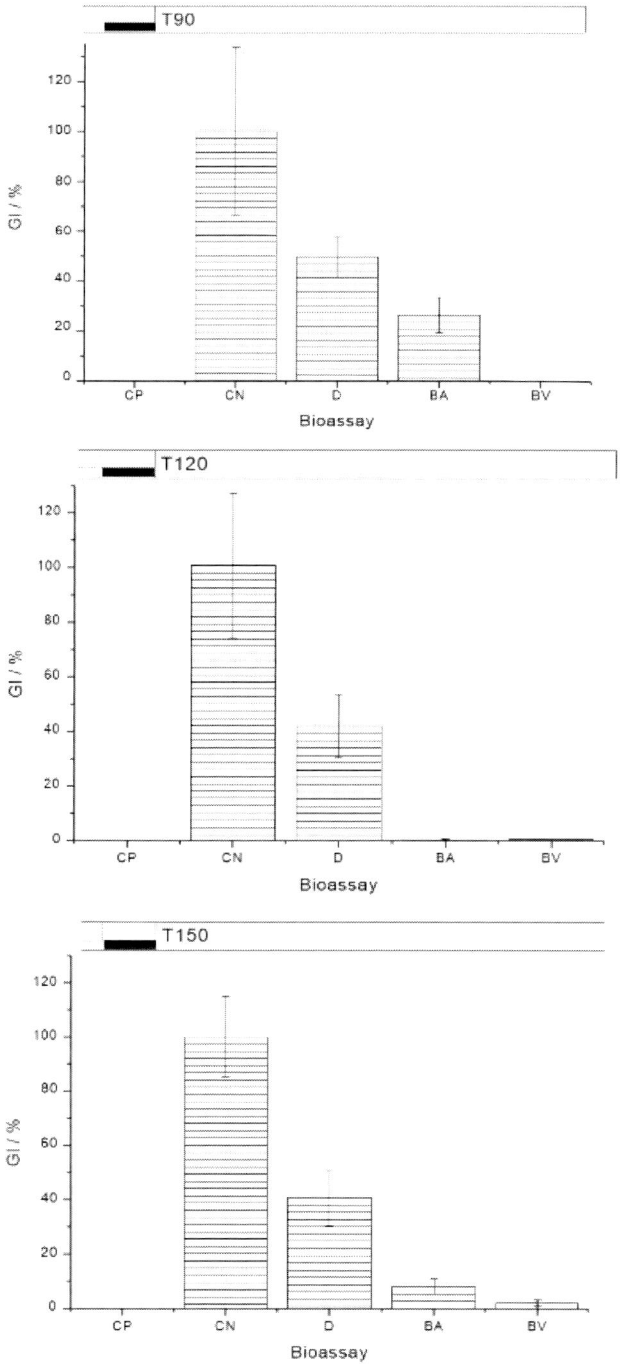

Figure 1. Germination of *Lactuca sativa* in bioassays at different biodegradation times.

Table 2. Root and hypocotil elongation of *E. sativa* seedlings

Bioassay	T0	T30	T60	T90	T120	T150
Hypocotyl	MeanStd. Dev.	MeanStd. Dev.	MeanStd. Dev.	MeanStd. Dev.	MeanStd. Dev.	MeanStd. Dev.
CN	2.7833 ± 0.6472	2.33 ± 1.1914	3.6467 ± 1.1328	2,505 ± 1,4268	1,8467 ± 1,466	1,465 ± 0,8967
D	1.495 ± 0.4987	0.925 ± 1.0801	2.4367 ± 0.5716b	1.55 ± 1.262	1.1217 ± 1.1367	0.9 ± 0.9015
BA	1.8533 ± 1.0272	1.2667 ± 1.1395	2.4867 ± 0.7777	1,366 ± 0,9951	0,0367 ± 0,2008	0,3783 ± 0,5771
BV	1.95 ± 0.8225	1.04 ± 1.0419	1.36 ± 1.2125	0.0717 ± 0.3925	0.335 ± 0.7144	0.1833 ± 0.4428
Root	MeanStd. Dev.	MeanStd. Dev.	MeanStd. Dev.	MeanStd. Dev.	MeanStd. Dev.	MeanStd. Dev.
CN	2.1833 ± 0.6126	2.3733 ± 1.1333	4.9717 ± 1.44	2,9667 ± 1,8482	2,2617 ± 1,4138	2,145 ± 1,0323
D	1.295 ± 0.3284	1.0783 ± 1.2218	5.075 ± 1.3831	2.0883 ± 1.8189	1.5333 ± 1.6604	1.2567 ± 1.1496
BA	1.26 ± 0.6333	1.3183 ± 1.1907	3.4633 ± 1.3309	0,9567 ± 0,7458	0,0433 ± 0,2373	0,4967 ± 0,7917
BV	1.2083 ± 0.501	1.3183 ± 1.3131	0.87 ± 0.8358	0.03 ± 0.1643	0.0483 ± 0.1004	0.2533 ± 0.5998

Table 3. Root and hypocotil elongation of *L. sativa* seedlings

Bioassay	T0	T30	T60	T90	T120	T150
Hypocotyl	MeanStd. Dev.	MeanStd. Dev.	MeanStd. Dev.	MeanStd. Dev.	MeanStd. Dev.	MeanStd. Dev.
CN	1.7567 ± 0.6239	1.0267 ± 0.656	2.4 ± 1.1921	1,4517 ± 0,8606	1,6817 ± 0,9229	1,2967 ± 0,7963
D	0.2983 ± 0.2571	0.11 ± 0.2053	1.0483 ± 0.7977	0.4917 ± 0.3917	1.3583 ± 1.0151	1.0217 ± 0.6741
BA	0.8667 ± 0.6063	0.25 ± 0.439	0.9083 ± 0.6369	0,53 ± 0,4401	0,0783 ± 0,1928	0,4017 ± 0,6521
BV	0.705 ± 0.5246	0.3333 ± 0.5964	0.4683±0.4403	0.026 ± 0.0828	0.01 ± 0.0548	0.1733 ± 0.3303
Root	MeanStd. Dev.	MeanStd. Dev.	MeanStd. Dev.	MeanStd. Dev.	MeanStd. Dev.	MeanStd. Dev.
CN	0.9483 ± 0.3861	1.295 ± 0.7357	2.6633 ± 1.4233	2,1267 ± 1,1172	1,6083 ± 0,9102	1,4167 ± 0,8209
D	0.3367 ± 0.3	0.21 ± 0.4444	2.11 ± 1.6249	0.6683 ± 0.5911	1.6067 ± 1.1884	1.305 ± 0.9479
BA	0.7817 ± 0.5037	0.345 ± 0.6828	1.5767 ± 1.4223	0,515 ± 0,5206	0,1067 ± 0,2599	0,495 ± 0,8051
BV	0.8433 ± 0.5536	0.4933 ± 0.7393	0.5367 ± 0.5372	0.03±0.0952	0.01 ± 0.0548	0.2267v0.4307

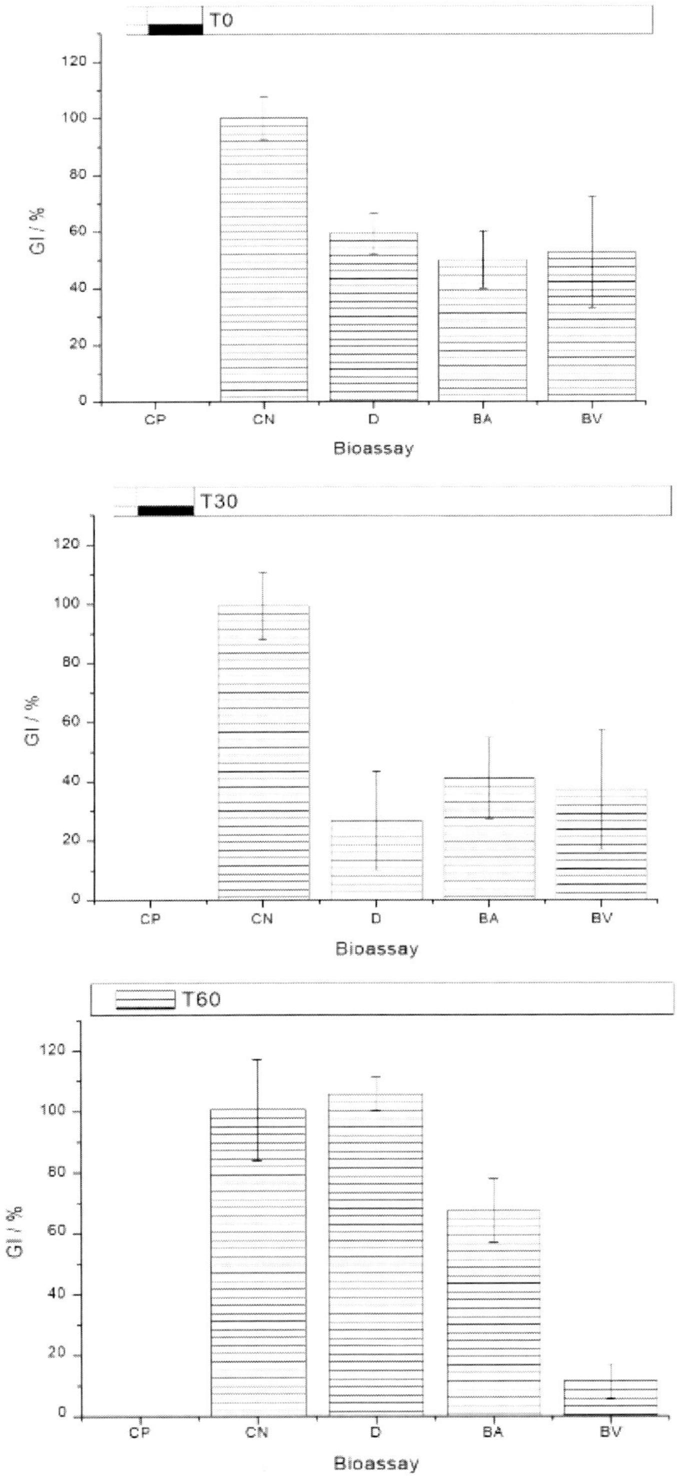

Figure 2. (Continued)

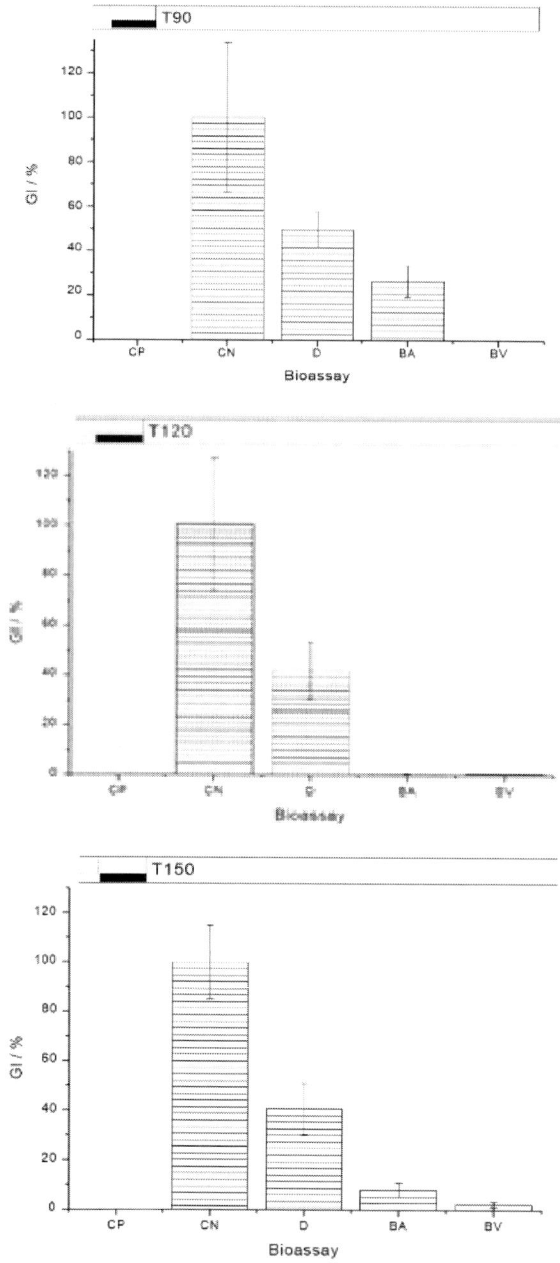

Figure 2. Germination of *Eruca sativa* in bioassays at different biodegradation times.

In T90 assays (Figure 1) the diesel assays showed an increase in toxicity (49.71% GI) and remained with values close to 40% in subsequent time points (T120 and T150).

Both bioassays containing biodiesel (from animal and vegetable sources) presented a toxicity increase across biodegradation time, presenting toxicity interference even after T120 (GI was lower than 10%). It has also been observed by Tamada et al., (2012) that seed germination of *E. sativa* and *L. sativa* in treatments with biodiesel in soil caused high inhibition according to the time of biodegradation in soil.

Table 2 shows the hypocotyl and root growth in bioassays using *E. sativa* seed. According to the Table 2, diesel bioassay causes less variation in the inhibition of *E. sativa* hypocotyl though the biodegradation process. The hypocotyls showed approximately 33%-60% growth inhibition at all times used. Biodiesel, however, did not show the same characteristics as diesel, since all samples varied widely their inhibition at different time points, with T120 and T150 presenting the highest inhibition compared to the negative controls.

From time T90 onwards, there is no longer a significant difference in vegetable tissue growth between the control and diesel bioassays. No difference was observed between the control and bioassays with biodiesel, either animal (BA) or vegetable (BV). There were, however, changes in root elongation at different biodegradation times.

The bioassays showed a slight variation in diesel inhibition at around 40% throughout almost all the analysis time. Surprisingly, at T60 a pattern change has occurred and root growth was larger than the control. In regards to biodiesel from animal source (BA), the toxicity was high at T120 and T150, with greater inhibition at the time T120. For biodiesel from vegetable source (BV) a more severe toxicity in T60 was observed, and its toxicity increased to nearly 100% inhibition at T90. High toxicity was kept up to the T150 time point.

Lactuca sativa Bioassays

According to Figure 2, changes occurred in the diesel (D) germination index (GI). The germination values were initially low, thus implying high toxicity in the T0 and T30. Along the diesel biodegradation process, GI increased reaching the values above negative controls, up to 100% in the final time points (T120 and T150). Hence, diesel toxicity through time decreases for *L. sativa*. However, contrary to diesel assays, tests with BA and BV were toxic over the analysis time. Toxicity was found particularly intense in BV, reaching GI values close to zero from the time T90.

Significant differences were found in biodiesel assays, presenting increased toxicity over biodegradation. The BV bioassay had specifically high inhibition values, especially after the T60 (Figure 2).

The elongation of *L. sativa* hypocotyl and root along the ground in biodegradation time relative to the negative control is presented in Table 3. There was a noticeable variation of the average growth of *L. sativa* seedling tissues. The diesel bioassay showed minor toxic response over time in relation to inhibition of hypocotyl growth. During the biodegradation process, diesel showed a decrease in the inhibition of hypocotyl reaching 21.21% inhibition compared to control at T150.

According to the average growth values, only diesel showed similarity to negative control assays and only at T120 and T150. There was a variation in growth elongation for diesel bioassays. Variations in plant growth inhibition occurred mostly during in early development times. From T60 onwards inhibition decreased and showed no significant difference from the controls. At T90 inhibition increases slightly, however, the final time points (up to T150) had its inhibition trend back to decrease, no significant difference in the inhibition of root bioassay with diesel compared to the negative control(Table 3).

Bioassays with biodiesel contaminants (BA and BV) initially shared lower toxicity values, evidenced by the difference in root growth medium (Table 3) compared to diesel. However, over time, the biodegradation of biodiesel bioassays proved to yield toxic compounds. The BV showed most intense toxicity of all substances.

Biodiesel Blends Phytotoxicity

In general, longer biodegradation times of biodiesel in soil increased toxicity to *E. sativa*. Diesel, however, was toxic in the early days of the bioassays. It may be related to its environmental behavior in soil. Adam and Duncan (2002) concluded that diesel involves the seeds thus creating a physical barrier preventing gas exchange and input water. In other words, diesel turns germination difficult by altering physiological parameters. Ogbo (2009) states that diesel may interfere with root growth. Such interference was found in our study. Significant differences occurred in the seedling growth pattern when compared to negative controls in some cases.

Vegetable biodiesel showed high inhibition, especially after 90 days of biodegradation. This shows that the biodiesel degradation considerably increases the toxicity of the polluted soil due to its by-products, as it forms peroxides and organic acids to reduce the pH of the soil, a fact observed by Bouaid et al., (2007) in his previous studies with vegetable biodiesel. The decrease in soil pH inhibits higher organisms, such as those used test organisms; in addition to interfering in the local microbiota, as these will only increase the biodegradation with the help of the surfactant.

An interesting fact was the high toxicity observed in bioassays containing biodiesel from animal or vegetable source, with no significant difference for all bioassays in the final days of biodegradation. The greatest toxicity was found in vegetable biodiesel, particularly in the final period of biodegradation in soil. The assays allowed inferring that the further biodegradation processes occur, the higher the toxicity of BA and BV contaminants.

References

Adam, G., Duncan, H., (2002). Influence of diesel fuel on seed germination. *Environ. Pollut.*, 120, 363-370.

Barros, S. B. M., Davino, S. C., (2003). *Avaliação da toxicidade.* In: Oga, S. Fundamentos de Toxicologia. São Paulo: Atheneu, 57-68.

Bouaid, A., Martinez, M., Aracil, J., (2007), Long storage stability of biodiesel from vegetable and used frying oils, *Fuel*, 86, 2596-2602.

Cetesb: Companhia de Tecnologia e Saneamento Ambiental, (1990). *Solos: Determinação de Resíduos: Método Respirométrico de Bartha.* Norma técnica L.6.350.

Ferrari, R. A., Souza, W. L., (2009), Avaliação da estabilidade oxidativa do biodiesel de óleo de girassol com antioxidantes. *Quim Nova*, 32, 106-111.

Geris, R., Santos, N. A. C., Amaral, B. A., Maia, I. S., Castro, V. D., Carvalho J. R. M., (2007), Biodiesel de Soja: Reação de Transesterificação para Aulas Práticas de Química Orgânica. *Quim Nova*, 30, 1369-1373.

IEA: International Energy Agency, (2011), Avalilable at http://www.iea.org/ stats/index.asp, Accessed in 09 set 2012.

Inazaki, T. H., Pião, A. C. S., Bidoia, E. D., Régis, G., Angelis, D. F., (2001).Teste de toxicidade utilizando sementes de *Eruca sativa. Arquivos do Instituto Biológico*, 68, 1-131.

Landis, W. G., Yu, M. H., (1998). *Introduction to environmental toxicology: Impacts of chemical upon ecological systems,* Florida, Lewis Publishers, p. 390.

Lopes, P. R. M., Bidoia, E. D., (2009) Evaluation of the biodegradation of different types of lubricant oils in liquid médium. *Braz Arch Biol Techn*, 52, 1285-1290.

Lopes, P. R. M., Montagnolli, R. N., Domingues, R. F., Bidoia, E. D., (2010).Toxicity and biodegradationin sandy soil contaminated by lubricant oil. *B Environ Contam Tox*, 84, 454-458.

Morales, G. C., (2004). *Ensayos Toxicológicos y Métodos de Evalución de Calidad de Agua: Estación, Intercalibración, Resultados y Aplicaciones.* México: IMTA, p. 189.

Oga, S., Siqueira, M. E. P. B., (2003), *Introdução à toxicologia.* In: Fundamentos de Toxicologia. São Paulo: Atheneu, 1-7.

Ogbo, E. M., (2009), Effects of diesel fuel contamination on seed germination of four crop plants: Arachis hypogaea, Vigna unguiculata, Sorghum bicolor and Zea mays, *Afr. J.Biotech.*, 8, 250-253.

Passarelli, M. M., (2003), *Poluentes da Atmosfera.* In: Oga, S. Fundamentos dc Toxicologia. São Paulo: Atheneu, 102-104.

Pinto, A. C., Guarieiro, L. L. N., Rezende, M. J. C., Ribeiro, N. M., Torresb, E. A., Lopes, W. A., Pereira, P. A. P., Andrade, J. B., (2005). Biodiesel: an overview. *J. Braz. Chem. Soc.*, 16, 1313-1330.

Suter, G., (1993), *Organism-level Effects,* In: Suter, G. W. I., Ecological risk assessment. Michigan: Lewis Publishers, p. 538.

Tamada, I. S., Montagnolli, R. N., Lopes, P. R. M., Bidoia, E. D., (2012).Toxicological evaluation of vegetable oils and biodiesel in soil during the biodegradation process. *Braz. J. Microb.* 43, 1576-1581.

Tiquia, S. M., Tam, N. F. Y., Hodgkiss, I. J., (1996), Effects of composting on phytotoxicity of spendpig-manure sawdust litter. *Environ. Pollut.* 93, 249-256.

Wang, W., (1987), Root elongation method for toxicity testing of organic and inorganic pollutants, *Environ Toxicol Chem*, 6, 409-414.

Zagatto, P. A., Bertoletti, E., (2006), *Ecotoxicologia Aquática: Princípios e Aplicações.* 1 ed. São Paulo, SP, Brasil. Editora Rima, p.478.

In: Soil Management
Editor: Manuel Esteban Lucas Borja

ISBN: 978-1-63483-274-8

Chapter 3

GEOTECHNOLOGY APPLIED TO ESTIMATE SOIL LOSS BY WATER EROSION

Francienne Gois Oliveira[1,*], Manuel Esteban Lucas Borja[2] and Odivaldo José Seraphim[3]

[1]Departament of Environmental Engineering,
University of Uberaba – Uniube/ Uberaba-MG, Brazil
[2]Department of Agroforestry Technology and Science and Genetics, School of Advanced Agricultural Engineering,
Castilla La Mancha University, Campus Albacete, Spain
[3]Department of Rural Engineering,
Universidade Estadual Paulista – UNESP/FCA – Botucatu/SP, Brazil

ABSTRACT

Water erosion is a major environmental problem that may compromise agricultural areas and the volume of water in a watershed. This paper presents the estimated soil loss in a watershed using Equation Revised Universal Soil Loss - RUSLE applied to a geographic information system. The factors of the RUSLE, rainfall erosivity (R), soil erodibility (K), topographic (LS), use factors and management conservation practices (CP) were obtained separately and used existing data that has been conducted in the region. In determining the potential erosion of Natural, we used the RKLS factors. All factors were generated and processed in a GIS. The results were obtained by means of maps and were satisfactory. Soil losses, predominately "Low" class, ranked among 0-10 t ha^{-}1 yr^{-}1. The GIS proved to be feasible in the determination of the various factors of Universal Soil Loss Equation, making possible zoning areas with high soil losses and high erosion potential.

Keywords: geographic information system, universal equation of soil loss, soil conservation

---

* Corresponding Author address: Email: franciennegois@yahoo.com.br.

INTRODUCTION

Water erosion is currently one of the main environmental problems for degrading soil and water resources. In addition, it poses a risk to food safety and represents a serious barrier to sustainable development (Teles et al.; 2011). Studies on soil use and management in watersheds, especially predictive models, are fundamental to reduce erosive processes. Models developed to calculate soil erosion can be divided into empirical, conceptual and those based on physical processes (Aksoy & Kavvas, 2005; Kinnel, 2010).

Empirical models such as the universal soil loss equation (Usle) (Wischmeier & Smith, 1978) and its revised version, Rusle (Renard et al., 1997), are used worldwide to provide useful information to support soil and water conservation plans (Kinnell, 2010; Oliveira et al., 2011). Annual soil loss is estimated from the combination of the six factors contained in Rusle: rainfall erosivity (R); soil erodibility (K); slope length (L); slope steepness (S); soil use and management (C); and support practices (P). According to Risse et al., (1993), topography factors (LS) and the C factor exerts the greatest influence on the global model efficiency.

The C factor corresponds to soil loss under specific cropping conditions, in relation to that occurring in bare soil (Wischmeier & Smith, 1978). It ranges from 0 (high plant cover) to 1 (bare soil). The Rusle uses five subfactors to calculate the C factor: residual effect of soil use (soil management); soil cover by plant canopy; soil cover by crop residues; roughness of soil surface; and soil moisture (Renard et al., 1997). The evaluation of each subfactor, however, is difficult because of the many possible combinations, and the time spent with data acquisition and analysis (Schönbrodt et al., 2010). Constant values of the Rusle C factor, produced in earlier studies, are usually used to evaluate soil erosion in watersheds. These values, however, do not accurately represent vegetation variation, particularly in large areas, which can result in mistaken estimates of soil loss (Asis & Omasa, 2007).

As you have all the predetermined parameters, the use of a geographic information system can contribute significantly to the spatial and mapping of areas with high rates of soil loss. This study aims to estimate and map the loss of water erosion of soils in a watershed applied to a geographic information system.

MATERIAL AND METHODS

Study Area

The study was conducted at the Lavapés Stream watershed, at Botucatu city, State of São Paulo, Brazil. The watershed is delimited from 22°43'12" S and 48°29'43"W. The watershed has an area of 44.01 km². The climate is defined as mesothermal climate, identified as Cwa (subtropical dry and hot summer inverse), according to Koppen. The vegetation is defined as cerrado and semideciduous forest ecosystem belonging to the Atlantic Forest biome (Embrapa, 2006).

Equation Universal Soil Loss

The equation of soil loss - RUSLE - was initially developed for use in agricultural and cultivated land, but was adapted for other types of use and occupation of land, including areas of urban construction, recreation sites, landfills, among others. The mathematical expression representing the ratio of the main factors influencing the erosion caused by rain, (Wischmeier and Smith, 1978; RENARD et al., 1997) as follows: A = R.K.L.S.C.P, in which: A is the annual mean soil loss (Ton. ha^{-1}. $year^{-1}$); R is the rainfall erosivity factor (MJ mm ha^{-1} $h-^{1}$); K is the soil erodibility factor (Ton. ha h MJ^{-1} mm^{-1}); L is the slope length factor, dimensionless; S is the slope steepness factor, dimensionless; C is the soil use and management factor, dimensionless; and P means the dimensionless factor for conservation practices.

Erosivity index (R) was obtained using the free software netErosividade SP, developed by the Research Group Water Resources, Federal University of Viçosa (GPRH / UFV) in partnership with the Agronomic Institute of Campinas (IAC) (MOREIRA et al., 2006), using the central geographical coordinates and the altitude of the watershed.

The average annual erosivity index of Lavapés Stream watershed was determined by NetErosividade SP program from the central geographic coordinates: 22 ° 49'40 "S, 48 ° 26'01" W and 706 meters altitude.

Soil erodibility (K factor) was estimated using the values found in Mannigel, et al., (2002) and Bertoni & Lombardi Neto (2012) that determined the erodibility factor and loss tolerance of São Paulo soils.

To obtain slope length (L) and slope steepness (S), a slope map and the digital elevation model was developed, the result was applied to the equation by Bertoni (1959): $LS = 0{,}000984.L^{0{,}63}.S^{1{,}18}$.

The factors CP (land use and conservation practices) were obtained by generating the map of land use classes where each class found was valued according to the methodology proposed by Stein et al., (1987).

Soil Loss was estimated through the Rusle model (Renard et al., 1997), multiplying the values of factors: RKLSCP obtained maps in shapefile format.

The resulting map was classified qualitatively according to the classification proposed by the United Nations Food and Agriculture (FAO, 1967), ranging from Low (0-10) to moderate (greater than 200) with values expressed in t.ha $^{-1}$.yr^{-1}. All factors were generated in map in shapefile format using a geographical information system free.

Results and Discussion

For the rainfall erosivity index (R) was against the value of 7.797 MJ.mm.ha-1.h-1.year-1, obtained through the NetErosividade / SP program, which uses the methodology proposed by Lombardi Neto and Moldenhauer (1992).

The soil erodibility (K) that is directly related to soil classes found in the study area, according to the Methodology of Mannigel et al., (2002) and Bertoni and Lombardi Neto (2012), this spatialized in Picture 1. Eigth soil classes were identified: Latossolos Vermelho-Amarelos Distróficos (LVAD), Latossolos Vermelhos Distróficos (LVd), Latossolos

Vermelhos Distroférricos (LVdf), Neossolos Flúvicos (RU), Neossolos Litólicos (NLd), Neossolos Litólicos Eutróficos (RLe) and Nitossolos Vermelhos (NVd).

The soil class - Latossolo Vermelho-Amarelo Distrófico is the most found in the study area (53.5%). This type of soil varies from deep to very deep with clayey to clayey. Normally relief plan and gently rolling, also occurring as undulating, strong wavy and mountainous. The low fertility inhibits the use of this type of soil for agriculture. Present silt content less than 20% and clay between 15% and 80%. Are soils with high water permeability, can be worked in wide range of moisture. Are deep, porous, well-drained, well permeable even when very friable clay and easy to prepare. Despite the high potential for agriculture, part of its area must be maintained to reserve for biodiversity protection of these environments (Embrapa, 2013).

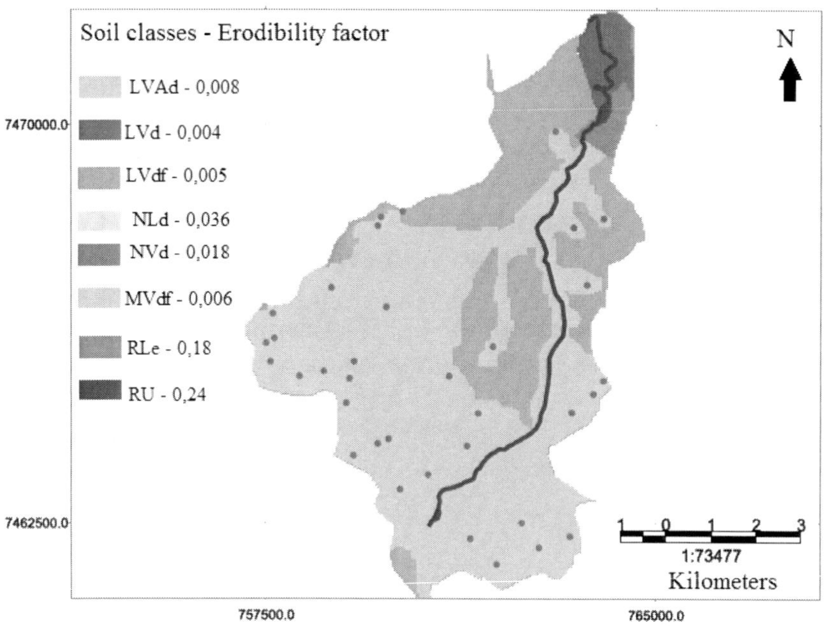

Picture 1. Soil map and K factor (amounts in $t.h.MJ^{-1}.mm^{-1}$).

The predominant relief is undulating (slope: 6-12%) to gently rolling (slope: 12-20%) according to Lepsch, et al., (1991). The variation of the LS factor was 0 to 27.57 that correspond directly to the slope classes. Most of the area showed 96.7% LS factor in the range of 0 - 10. In between 10 and 27.57 is approximately 1.5 square kilometer area (Picture 2).

The CP factors are directly related to the use and land cover. Were determined using 4 classes: building area - 71.5%, field / pasture - 18.2%, forest - 8.8% and water - 1.5%. The class that most exposes the soil to erosion covers the field area / pasture. In Figure 2 the CP values are spatially and land use classes.

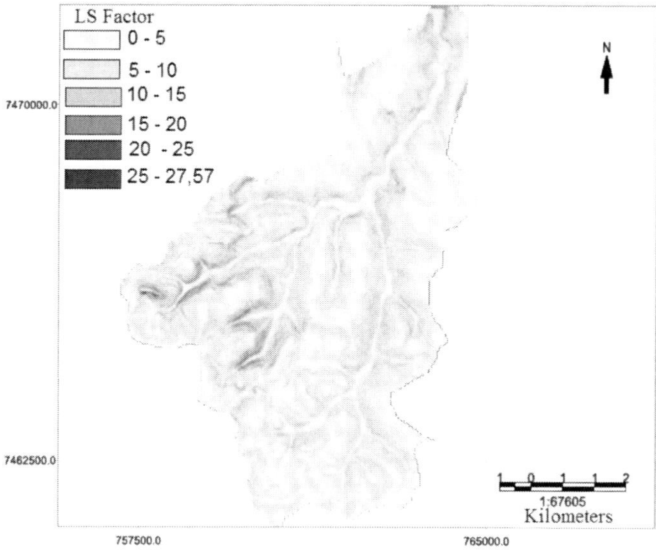

Picture 2. Classes LS factor.

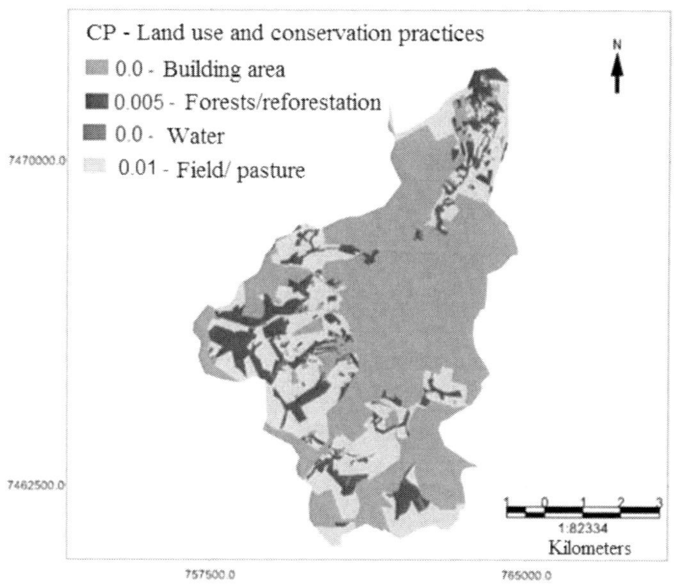

Picture 3. Land use map and CP factor of Lavapés Stream watershed with amounts in $t.ha^{-1}.yr^{-1}$.

Regarding the estimation of soil loss is observed that the spatial distribution of susceptibility to erosion is associated with relief and land use classes. Areas with estimates exceeding 50 $t\ ha^{-1}.year^{-1}$ are located in regions of maximum gradient and classified as field / pasture.

Soil losses concentrated among 0-10 $t.ha^{-1}.year^{-1}$ are seen in areas where the CP factor ranges from 0 to 0.005 (area built and forests / reforestation) and are classified as low. Some small areas with loss of soil classified too high are observed in the Neossolo Litólico soil. This type of soil has a high susceptibility to erosion.

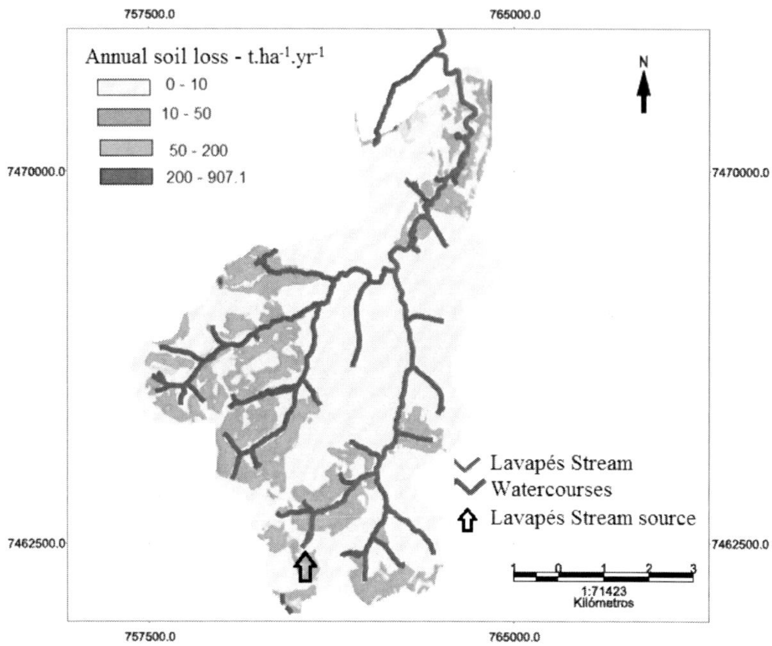

Picture 4. Geographical distribution of the estimated soil loss by erosion in the Lavapés stream watershed as amounts in t.ha^{-1}.yr^{-1}.

CONCLUSION

The results achieved with the use of the Universal Soil Loss Equation in the estimated soil losses in the Lavapés stream watershed are satisfactory. Soil losses were predominance low, classified 0-10 t.ha^{-1}.yr^{-1}.

The free geographic information system is feasible in determining the different factors of Equation Universal Soil Loss. It was possible to identify and quantify the areas with high losses of soil. From the data obtained, it is possible to draw management plans and erosion control watershed.

ACKNOWLEDGMENTS

To Coordenação de Aperfeiçoamento de Pessoal de Nível Superior (Capes) and Conselho Nacional de Desenvolvimento Científico e Internacional (CNPq) for scholarship granted.

REFERENCES

Aksoy, H., Kavvas, M. L. 2005. A review of hillslope and watershed scale erosion and sediment transport models. *Catena*.64, 247-271.

Asis, A. M. de, Omasa, K. 2007. Estimation of vegetation parameter for modeling soil erosion using linear spectral mixture analysis of Landsat ETM data. ISPRS Journal of Photogrammetry and Remote Sensing. 62, 309-324. DOI: 10.1016/j.isprsjprs.2007.05.013.

Baptista, G. M. M. 2003. Diagnóstico Ambiental de Erosão Laminar: modelo geotecnológico e aplicação. Brasilia.

Bertoni, J., Lombardi Neto, F.2012. Conservação do solo. 8. São Paulo.

Bertoni, J.1959. O espaçamento os terraços em culturas anuais, determinados em função das perdas por erosão. Bragantia: Boletim técnico do instituto agronômico do Estado de São Paulo. 18,113-140.

Empresa Brasileira de Pesquisa Agropecuária. 2006. Sistema brasileiro de classificação de solos. Embrapa Solos. 2. Rio de Janeiro.

Empresa Brasileira de Pesquisa Agropecuária. 2013. Embrapa Solos. Latossolo vermelho-amarelo. 2. Rio de Janeiro.

Food and Agriculture Organization. 1967. La erosión del suelo por el água. Algunas medidas para combatirla en las tierras de cultivo. Cuadernos de Fomento Agropecuário da Organización de las Naciones Unidas. Roma: FAO. 81, 207.

Kinnell, P. I. A. 2010. Event soil loss, runoff and the Universal Soil Loss Equation family of models: a review. *Journal of Hydrology*. 385, 384-397, 2010.

Lepsch, I. F., Bellinazzi Jr., R., Bertolini, D., Espíndola, C. R. 1991. Manual para levantamento utilitário do meio físico e classificação de terras no sistema de capacidade de uso. 4ª aproximação. Campinas.

Lombardi Neto, F., Moldenhauer, W. C.1992, Erosividade da chuva: sua distribuição e relação com as perdas de solo em Campinas (SP). Bragantia, Campinas. 51, 189-196.

Mannigel, A. R. et al. 2002. Fator erodibilidade e tolerância de perda dos solos do Estado de São Paulo. Acta Scientiarum. 24, 1335-1340.

Moreira, M. C. et al. 2006. Programa computacional para estimativa da erosividade da chuva no estado de São Paulo utilizando redes neurais artificiais. Engenharia na Agricultura, Viçosa, MG. 14, 88-92.

Oliveira, P.T.S., Alves Sobrinho, T.. Rodrigues, D.B.B., Panachuki, E. 2011. Zoneamento ambiental aplicado à conservação do solo e da água. Revista Brasileira de Ciência do Solo, 35, 1723-1734. DOI: 10.1590/S0100-06832011000500027.

Renard, K. G., Foster, G. R., Weesies, G. A., Mccool, D. K., Yoder, D. C. 1997. Predicting soil erosion by water: a guide to conservation planning with the Revised Universal Soil Loss Equation (RUSLE). Agriculture handbook, 703, Washington

Risse, L.M., Nearing, M.A., Nicks, A.D., Laflen, J.M.1993. Error assessment in the Universal Soil Loss Equation. *Soil Science Society of America Journal*.57, 825-833. DOI: 10.2136/sssaj1993.03615995005700030032x.

Schönbrodt, S., Saumer, P., Behrens, T., Seeber, C., Scholten, T. 2010. Assessing the USLE crop and management factor C for soil erosion modeling in a large, mountainous watershed in Central China. *Journal of Earth Science*. 21, 835-845. DOI: 10.1007/s12583-010-0135-8.

Stein, D. P., Donzelli, P. L, Gimenez, A. F., Ponçano, W. L., Lombardi Neto, F. 1987. Potencial de erosão laminar, natural e antrópico, na bacia do Peixe – Paranapanema. Simpósio nacional de controle de erosão. 3, 105-135.

Telles, T. S., Guimaraes, M. F., Dechen, S. C. F., 2011. The costs of soil erosion. *Rev. Bras. Ciênc. Solo*. 287-298.

Wischmeier, W. H., Smith, D. D. 1978. Predicting rainfall erosion losses: a guide to conservation planning. Agriculture handbook, 537, Washington.

In: Soil Management
Editor: Manuel Esteban Lucas Borja
ISBN: 978-1-63483-274-8

Chapter 4

SOIL REMEDIATION TECHNOLOGIES: REHABILITATING DEPLETED AND POLLUTED SOILS IN SUB-SAHARAN AFRICA

***Kofi Gyan Quartey*[1,*] *and Georgina Dede Arthur*[2]**
[1]School of Social Sciences, University of KwaZulu-Natal,
Pietermaritzburg, South Africa
[2]Department of Nature Conservation, Mangosuthu University of Technology, Durban
KwaZulu-Natal, South Africa

ABSTRACT

Soil is the foundation of basic ecosystem which remains important for the sustenance of plants and animals. In less developed countries, agrarian land forms an integral part of livelihood. In recent times, soils in Sub-Saharan Africa have been subjected to various natural and anthropogenic factors that have contributed to their pollution and degradation. For instance, climate change has rendered large hectares of lands unproductive while soil contaminations from industrial and agricultural chemicals have also led to 'sick' soils. This paper investigates the types of soil remediation technologies (SRT) that are suitable for rehabilitating degraded and polluted soils in Sub-Saharan Africa.

Keywords: climatic change, desertification, soil pollution, food security, remediation

INTRODUCTION

Soil contamination is considered a serious problem because it affects the environment and living organisms as well as the economic activities associated with the use of soil (Wania, 1999). Soils are the uppermost layer of the earth's crust and one of the most important non-renewable resources (Blum et al., 2006). Soil is the unconsolidated mineral or organic matter

[*] Corresponding author: Email: Georgina@mut ac.za. Tel: +27 31 907 7622, +27 724 732 437, Fax +27 862 737.

on the surface of the earth that has been subjected to and shows effects of genetic and environmental factors (SSSA, 1997). The unconsolidated materials that make up soil consist of minerals from weathered rocks, dead, decaying organic matter, water, air and living organisms. Healthy soil contributes to the significant production of biomass, storage, filtration and transformation of water, nitrogen and carbon (Treville, 2011). The potential of soil to sustain agricultural productivity has been declining over the years. Currently, about 60% of the population of Sub-Saharan Africa depends on agriculture for their livelihood (Molden and de Fraiture, 2004). Thus, it is imperative that the sustainable management of the soil is ensured (Lal, 2007). With the dawn of the industrial revolution, the unprecedented increase in industrial activities coupled with increasing human population and increased urbanization has resulted aggravated the strains on soil and water resources (Daverey and Pakshirajan, 2011). Contamination of soils by hazardous substances is a major environmental concern worldwide. Generally, Sub-Saharan Africa is heavily dependent on functioning soils for their commercial and subsistence agricultural needs. Soil remediation techniques are fast growing into vital options in modern environmental friendly economy. Reclamation techniques are aimed at restoring land to beneficial use or to reduce the contamination in the soil (Okeke, 2013). Soil remediation involves various techniques useful for the management of degraded sites. It involves clean-up of contaminated sites using physical, chemical and biological techniques which could be either *in-situ* or *ex-situ*. While *in situ* soil remediation techniques involve the treatment of contaminated materials at source, *ex situ* involves the transfer or removal of the contaminated materials for treatment elsewhere (Boopathy, 2000). The increase rate of soil pollution has led many researchers to focus on developing

fast, economical and more efficient remediation technologies (Karami and Shamsuddin, 2010). There exists a wide range of physio-chemical treatment processes available for soil remediation including bioremediation methods. In order to decontaminate soils, the primary contaminants have to be detected and quantified (Pavel and Gavrilescu, 2008). This chapter gives an overview of the importance of soil and highlights various sources of soil contamination. In addition, the various *in situ* and *ex situ* approaches that include bioremediation technologies are also discussed.

2. Soil

2.1. Soil and Its Importance to Biodiversity

Soil has a role as a habitat and serves as a platform for diverse human activities (Usman, 2012). Soils are made up of vertical columns of unconsolidated matter, commonly referred to as horizons (Figure. 1). These are made up of mineral and organic constituents, water and gases that are characterized by physical properties and chemical as well as physical-chemical properties which distinguishes them from one another and from the parent material.

Soils are naturally sensitive as the rate at which it forms and regenerates is relatively slower process when compared to its degradation (Juana 2007). According to FAO (2014), it can take up to 1,000 years for the earth to form one centimeter of soil. Currently in Sub-Saharan Africa, extensive biochar projects have been implemented to sequester CO_2 from the climate change mitigation and enhance soil quality for increased food security (Whitman and Lehmann, 2009; Lal, 2010; Smith et al., 2014). Declining soil quality can be attributed to modern society's increased demands for water, energy, and land areas for activities such as agriculture, urbanization, disposal of industrial and urban wastes (Lal, 2007). Based on the suggestion of FAO's Global Soil Partnership and with the support of the EU, in December 2013 the General Assembly of the United Nations proclaimed 2015 as the 'International Year of Soils' (IYS) and 5th December of each year as the 'World Soil Day' (WSD) (Environment, 2015).

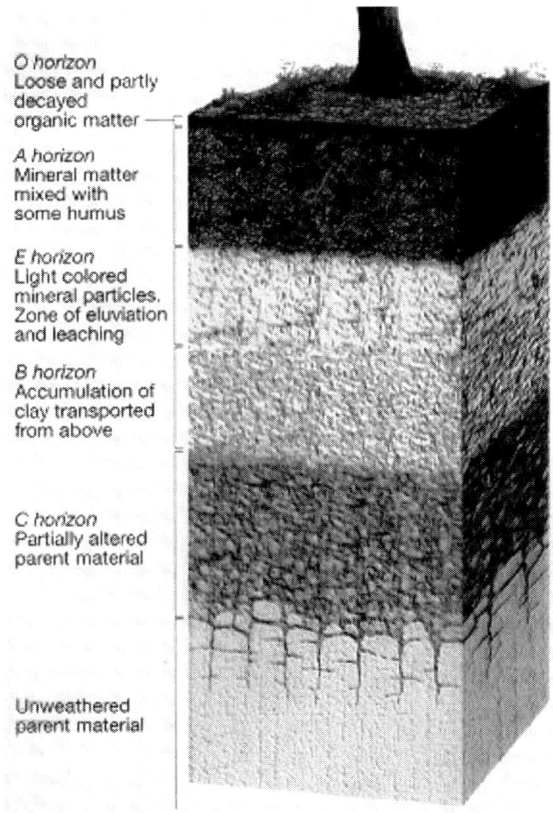

Source: http://courses.missouristate.edu/ejm893f/creative/glg110/Weathering.html.

Figure 1. Soil Horizon A portrayal of the horizons within the profile of a typical forest soil.

3. Man-Made Sources Of Soil Degradation And Contamination

Soil being a "universal sink" bears the greatest burden of environmental pollution. The two main sources of soil pollution are agricultural and non-agricultural bases. Figure 2 shows the diverse sources for the soil pollution including mining exploitations and waste dumping (Ashraf et al., 2014). It is considered a serious problem as it affects living organisms and human health as well as the economic activities associated with the use of soil (Wania, 1999).

Soil degradation resulting from pollution affects agriculture productivity, wellbeing of rural and commercial farmers in most Sub-Saharan African countries. Soil degradation occurs when the current and potential ability of the soil to produce goods and services quantitatively and qualitatively is reduced (FAO-UNEP, 1983). Contamination leads to land degradation which causes a loss of ecosystem function and services from which the system cannot recover unassisted. Oil refineries, gas stations, use of wood preservatives, agrochemicals, petrochemical and pharmaceutical are main sources of toxic contaminants (Nedunuri et al., 1998; Mrozik, 2010). Major sources of heavy metals in soils are the result of other anthropogenic activities such as active and abandoned industrial activities, improper waste dumping and automobile emissions (Azeez et al., 2011).

3.1. Abandoned Industrial Activities

Mining has contributed to the growth of the economies of many Sub-Saharan African countries. Unfortunately, the activity of mining on the Sub-continent has also contributed to increased soil pollution (Akiwumi and Butler, 2008). For instance, soil samples collected from Zambia near Kwabe abandoned (lead and zinc) mines showed high levels of soil contamination such as Pb, Zn, Cu, Cd, and As (Nakayama et al., 2011).

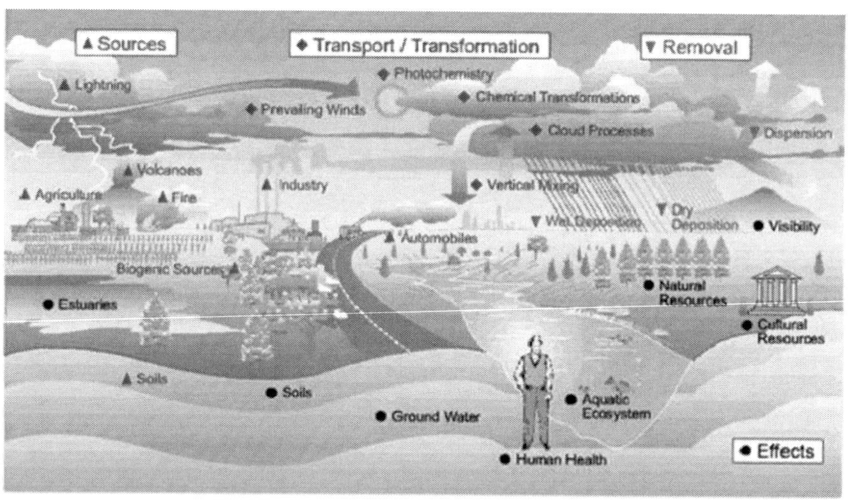

Figure 2. Sources of soil pollution (Ashraf, 2014) accessed 23rd February 2015).

3.2. Automobile Emissions

Exposure to ambient fine particular matter (PM2.5) was responsible for 3.2 million premature deaths in 2010 and is among the top ten leading risk factors for death. Surface transportation is a significant global source of PM2.5 emissions and a target for new actions (Chambliss et al., 2014). The available information suggests that the concentrations of toxic metals in many ecosystems of Africa are reaching unprecedented levels. Due to heavy load of contaminated dusts in the atmosphere of the overcrowded cities, the ambient concentrations of toxic metals are currently among the highest being reported anywhere. Lead pollution from the increasing number of automobiles and cottage industries represents a major health hazard, and it is estimated that 15–30% of the infants in some urban areas may already be suffering from lead poisoning (Nriagu, 1992).These types of pollutions include emission of metals from vehicle exhausts disposition on nearby roadside soils (Ghrefat and Yusuf, 2006; Panichayapichet et al., 2007). Highway traffic also generates heavy metals and particulate matter through various vehicular and tire-pavement abrasion mechanisms (Azeez et al., 2009). These abraded materials, when deposited on the soil, accumulate and can prove a health hazard over time. Results from a study conducted in Abeokuta (Nigeria) to analyze the distribution of heavy metals in the soil indicates that emissions from vehicles remain the major source of these metals (Azeez et al., 2009). Similar observation was also reported in Botswana (Mmolawa et al., 2011).

4. Soil Remediation Technologies

In the last few years, research has investigated soil remediation technologies. Currently, there exists a wide range of contaminated soil remediation technologies. For instance, chemical remediation involves the use of chemicals for the treatment of contaminants. These chemicals grouped into three categories include dispersants, solidifiers and chemical oxidants (Dave and Ghaly, 2011; Abha and Singh, 2012). Other technologies are land filling, vitrification, electrokinetic, bioremediation and possible combined technologies as bioelectroremediation. According to Reddy et al., (1999), the application of these technologies is dependent on the characteristics of the soil such as the degree of the contaminant, economic viability and the estimated time duration. Remediation is often preferred over methods of excavation and removal, containment or covering re-vegetation because they are perceived as having added environmental value (Hazen, 2010). Generally, remediation technology can be categorized into non-biological, biological and combined remediation technologies.

4.1. Non Biological Remediation Technology

4.1.1. In Situ Vitrification

In situ vitrification (ISV) involves the passage of low level direct current between two electrodes in the soil to remove contaminants. The apparatus consist of four principal parts; soil cell, electrode compartments, electrolyte solution reservoir and power supply unit (Fig. 3). The electrical power heats and melts soil, sludge, mine tailings, buried waste and sediments contaminated with organic, inorganic and metal bearing hazardous waste. The molten material cools to form a hard monolithic, chemically inert, stable glass and crystalline product that incorporates and immobilizes the thermally stable inorganic compounds. The sludge product material is glass-like with very low leaching characteristics. Organic waste are initially vaporized or paralyzed by the process. These contaminants migrate to the surface where the majority is then burned within a hood covering. The residue treatment areas are treated in an off gas treatment system. ISV has been reported in treating a large variety of organic and inorganic waste based on the result of pilot scale test (EPA, 1994).

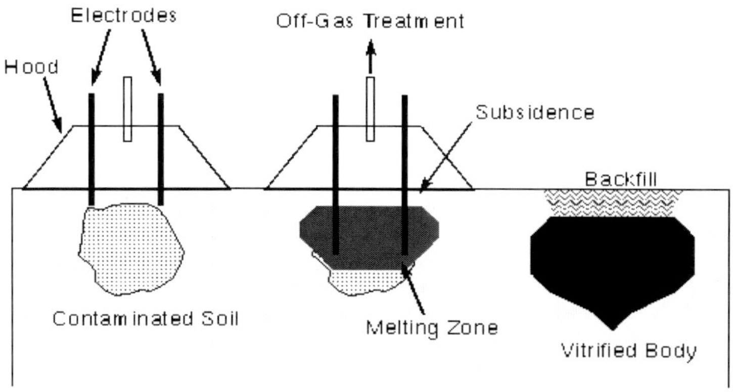

Figure 3. Stages of the in situ vitrification (ISV) process (http://www.math.utk.edu/~vasili/va/descr/isv/fig1.isv-drwg.gif.

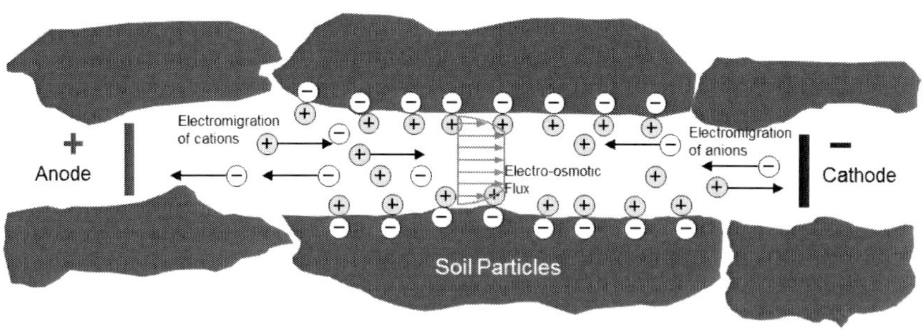

Figure 4. Transport mechanisms in electrokinetic remediation.

4.1.2. Electrokinetics

Electrokinetic technique involves the use of direct current to remove organic, inorganic and heavy metal particles from the soil by electric potential (Acar et al., 1993; Lyer, 2001; Wada et al., 2001). Some of the advantages of this technique are that remediation can be conducted to treat contaminants in low permeability zones to overcome accessibility of contaminants or delivery of treatment (Ho et al., 1995); remediation can occur despite soil stratification or homogeneity (Sharma, 2004). In addition, the technique has low cost of operation and potential applicability to a wide range of contaminant types (kin et al., 2000). According to Acar et al., (1995) and Virkutyte et al., (2002), electrokinetic technique has been applied in the treatment of soils contaminated with mercury, zinc, iron, lead, copper and manganese (Okeke 2013). Electrokinetic technique effectively reduced (from 38.45 mg/kg to 9.16 mg/kg translating to 76.2% reduction) the level of Cd concentration in Cd-contaminated soil. In addition, the procedure is environmentally friendly and relatively faster than bioremediation. According to EPA (2004), the technique has an advantage of cost effectiveness and its ability to leave the soil in its original conditions, as an in-situ technique (Figure. 4). It is gradually being recognized as an emerging solution to soils contaminated with heavy metals.

4.2. Biological Remediation Technology

Bioremediation is a sustainable alternative to chemical-physical processes and is attractive solution for soil decontamination and renewal of polluted sites (Tomei and Daugulis, 2013). It is a technique whereby microorganisms are stimulated to rapidly remove or degrade hazardous pollutants in soils (Pandey and Fulekar, 2012). Types of bioremediation related methods include composting, land farming, phytoremediation, bioventing, bioreactor, bioleaching, biostimulation and bioaugmentation,

4.2.1. Composting

Composting is a thermophilic treatment process in which contaminated material is mixed with a bulking agent using either static or aerated piles (Boopathy, 2000). Compost bioremediation is carried out by co-composting the contaminants with suitable compost materials to induce the biodegradation of the contaminant (Reid et al., 2000).

The principle of operation consists of mixing the contaminated soil with non-hazardous organic amendments, generally other solid wastes (for instance, manure and agricultural wastes) suitable for composting applications, to boost the bacterial populations required to degrade the pollutants in the soil (Tomei and Daugulis. 2015).

There are studies on degradation of organic pollutants using composts (Potter et al., 1999; Reid et al., 2000; Marin et al., 2006). Co-composting may be suitable for practical field application for effective in situ and ex-situ bioremediation of oil sludge. The process is an effective and controlled technology (with attributes such as nutrients, temperature, moisture, large population of microbes) for the degradation of oil sludge (Ubani, 2006). The use of vermicompost-amended soil would be effective to remediate the heavy metals from contaminated environment (Jadia and Fulekar, 2008). In principle, the vermicompost binds metals and increase uptake by providing nutrients such as sodium, magnesium, iron, zinc, manganese and copper. These serve as a natural fertilizer giving high yield of biomass and microbial consortium which helped the overall growth of the plant. Other organic amendments with potential are seaweed concentrates and compost tea which could be used singly or as mixtures.

4.2.2. Land Farming

Land farming is an aboveground remediation technology that reduces the concentration of contaminants in soils through processes associated with bioremediation (Khan, 2004). Land farming is a solid-phase treatment system for contaminated soils via either in situ or ex situ (Boopathy, 2000). It is one of the most widely used soil remediation technologies due to its low technological footprint and relatively low cost (Tomei and Daugulis, 2013). Globally, the technique has been practiced for over 100 years (Riser-Roberts, 1992). In situ bioremediation of petroleum waste sludge in land farming sites of motor oil Hellas (petroleum refinery) was studied by monitoring the changes of the petroleum composition of the waste sludge and the structure of the microbial community. Findings of the study indicated that the degradation of petroleum hydrocarbons was enhanced in the land farming areas and a depletion of n-alkanes of approximately 75–100% was achieved. During the Gulf Crisis, Kuwait was subjected to hazards, caused by the oil well fires and 300 oil lakes were formed covering land areas of exceeding 49 km^2 (Al-Awadih et al., 1996). Bioremediation techniques involving enhanced land farming was selected and evaluated at pilot scale to remediate the extensive oil spills at the Burgan oil field in which 16-land farming plots of 120 m^2 each were constructed. The land farming treatment resulted in more than 80% reduction of oil contaminants within 15 months. The treatment also resulted in a substantial reduction of the polycyclic aromatic hydrocarbon concentrations (Al-Awadhi et al., 1996).

4.2.3. Phytoremediation

The importance of biodiversity is increasingly considered for the clean-up of the metal contaminated and polluted ecosystems. Unlike organic compounds, metals cannot be degraded and their clean-up requires their immobilization and toxicity level reduction or complete removal. In recent years, researchers have generated cost effective technologies which includes use of microorganisms or plants for cleaning-up polluted areas (Kuzovkina et al., 2004; Qui et al., 2006). This subject is emerging as a cutting edge area of research gaining commercial significance in the contemporary field of environmental biotechnology (Prasad and Freitas, 2003). Phytoremediation, a cost-effective plant-based approach to mitigation takes advantage of the ability of plants to extract concentrated metals from the environment (Salt et al., 1988). According to Evangelou et al., (2015), the method may not be suitable where contamination of pollutants is too high and toxic. GWRTAC, (1996) as cited in Khan, Husain and Hejazi (2004) gives a brief account of the five basic types of phytoremediation: (1) rhizofiltration, water remediation techniques in which contaminants are taken up by the plant's roots; (2) phytoextraction, which involves the uptake of contaminant from the soil; (3) phytotransformation, which is applicable to both soil and water and involves the degradation of contaminants through plant metabolism; (4) phytostimulation or plant-assisted bioremediation, which involves the stimulation of microbial degradation through the activities of plants in the root zone; and (5) phytostabilization, which uses plants to reduce the migration of contaminants through the soil. Phytovolatilization where metal contaminants

such as As, Hg, and Se may exist as gaseous species in environment. Naturally occurring or genetically modified plants are capable of absorbing elemental forms of these metals from the soil, biologically converting them to gaseous species within the plant prior to release into the atmosphere (Watanabe, 1997). Phytoremediative technologies which are soil-focused are suitable for large areas that have been contaminated with low to moderate levels of contaminants. Sites which are heavily contaminated cannot be cleaned through phytoremediative due to the harsh conditions will not support plant growth (Schnoor et al., 1995). Hence organic amendments become useful in situation of this type. Table 2 indicates the types of plants for phytoremediation.

4.2.3.1. Rhizofiltration

Rhizofiltration or phytofiltration is an adsorption subset technique of phytoremediation whereby terrestrial and aquatic plants are utilized to remediate metal-contaminated groundwater, and also contaminated soils (Mulligan et al., 2001; Yadav et al., 2011). In the case of soil remediation the plants absorb concentrate and precipitate mater through their roots (Dushenkov et al., 1995). The metal enriched plants are then harvested in order to remove the metal contaminants that are absorbed in the root zone (Yadav et al., 2011). This method is an in situ approach that is highly cost effective and environmentally friendly

(Wantanabe, 1997). As shown in Table 1, the method has been used for the effective reduction of ^{137}Cs, ^{90}Sr and U (Schnoor, 1997).

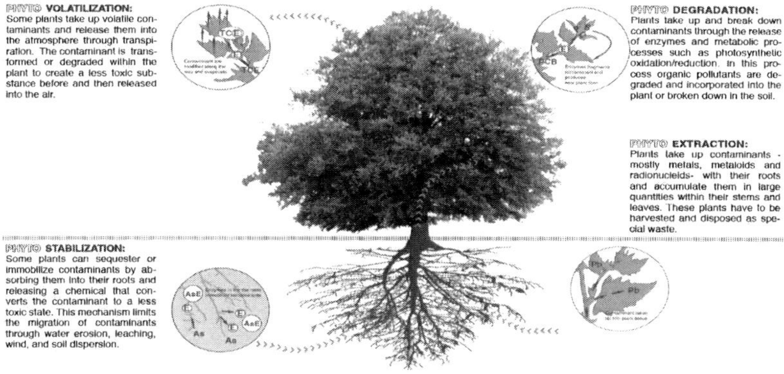

Figure 4. Different types of phytoremediation brownfields-to-greenfields-a-field-guide-to-phytoremediation Kühl (2010).

Table 1. Rhizofiltration technique to remediate nuclear disaster and energy wastes (Schnoor, 1997)

Location	Application	Plants	Performance	Reference
Chernobyl, Ukraine	Rhizofiltration near nuclear disaster	Sunflowers (*Helianthus anuus*)	90% reduction ^{137}Cs, ^{90}Sr in 2 weeks, 8000 X normal concentration in roots	Rutgers University
Ashatabula, OH	Rhizofiltration of energy wastes	Sunflowers (*Helianthus anuus*)	95% removal of U in 24 h (350 ppb to <5 ppb)	Phytotech

Table 2. Typical plants used in various phytoremediation applications (Schnoor, 1997)

Application	Media	Contaminants	Typical plants
Phytotransformation	Soil, ground water, landfill leacheate, land application of wastewater	Herbicides, aromatics, chlorinated aliphatics, nutrients, ammunition wastes	Phreatophyte trees (poplar, willow, cottonwood, aspen, Grasses (yre, bermuda, sorghum, fescue, Legumes (clover, alfalfa, cowpeas)
Rhizosphere bioremediation	Soil, sediments, land application of waste water	Organic contaminants (pesticides, aromatics and polynuclear aromatic hydrocarbons)	Phenolics releasers (mulberry, apple, osage orange) Grasses with fibrous roots (rye, fescue, bermuda) for contaminats 0-3 ft deep Phreatophyte trees for 0-10 ft Aquatic plants for sediments
Phytostabilization	Soil, sediments	Metals (Pb, Cd, Zn, As, Cu, Cr, Se, U), hydrophobic organics) PAHs, PCNBs, dioxins furans, pentachlorophenol, DDT, deldrin)	Phreatophyte trees to transpire large amounts of water for hydraulic control Grasses with fibrous roots to stabilize soil erosion Dense root systems are needed to sorb/bind contaminants
Phytoextraction	Soil, brown fields, sediments	Metals (Pb, Cd, Zn, Ni, Cu) with EDTA addition for Pb Selenium (volatilization)	Sunflowers Indian mustard Rape seed plants Barley, Hops Crucifers Serpentine plants Nettles, Dandelions
Rhizofiltration	Groundwater, water and wastewater in lagoons or created wetlands	Metals (Pb, Cd, Zn, Ni, Cu), radionuclides (137Cs, 90Sr, U), hydrophobic organics	Aquatic plants: Emmergents (bulrush, cattail, coontail, pondweed, arrowroot, duckweed); submergents (algae, stonewort, parrot, feather, Eurasian water, milfoil, Hydrilla

4.2.4. Bioventing

Bioventing involve treating contaminated soils by pumping oxygen through soil to stimulate microbial activity to metabolize fuel components (Dupont, 1993). This process injects air into the contaminated media at a rate designed to maximize *in situ* biodegradation and decrease or eliminate the off gassing of volatilized contaminants into the atmosphere (Lee and Swindoll, 1993; Khan, 2004). This method is relatively economical and has limited equipment requirements. It is non-invasive and causes minimal disturbances during operation (King et al., 2014). Bioventing offers a remedy for large areas of contaminated soil without requiring demolition of basements (Malina et al., 1999). Thus, bioventing is frequently proposed as an alternative to the commonly used remedy of excavation and surface treatment or landfilling of contaminated soils. Bioventing techniques have been successfully used to remediate soils contaminated by petroleum hydrocarbons, non-chlorinated solvents, some pesticides, wood preservatives, and other organic chemicals (Katsivela et al., 2005). Findings by Österreicher-Cunha et al., (2004) indicated that bioventing appears as an interesting tool in regenerating gasoline-ethanol contaminated soils because it accelerates soil detoxification and microbiota recovery.

4.2.5. Slurry Bioreactor

A slurry bioreactor is a containment vessel and apparatus used to create a three-phase (solid, liquid, and gas) mixing condition to increase the bioremediation rate of soil-bound and water-soluble pollutants (Vidali, 2001). Slurry bioreactors are one of the most significant types of *ad situ* and *ex situ* technologies that can be used for bioremediation of soils with high clay and organic matter contents with high pollution levels (Robles-González et al., 2012). According to Khan, Husain and Hejazi, (2004), this method requires excavation of the contaminated soil, which is treated in a controlled bioreactor. The excavated soil is mixed with water and other additives, along with selected bacteria. Treatment occurs at rapid rates ranging from 1-6 months. Bioslurry/bioreactors are successful in treating non-halogenated semi volatile organic compounds (SVOCs) and volatile organic compounds (VOCs) in excavated soils or dredged sediments, pesticides and PCBs (Khan, Husain and Hejazi, 2004). Success in remediating soils with toxic chemicals such as lindane (also known as hexachlorocyclohexane - HCH) which is widely used in agriculture and public health and has long been associated with serious soil pollution (Robles-González et al., 2012).

4.2.6. Bioleaching

Bioleaching involves the dissolution of metals from their mineral sources by natural organisms or the use of microorganisms to transform elements for extraction from a material when water is filtered through it (Mishra, 2005). Rohwerder et al., (2004) defines bioleaching as the mobilization of metal cations from insoluble ores by biological oxidation and complexation processes. Currently, a bioleaching offers attractive features for the extraction of metals from solid materials, such as lower cost and energy requirements, environmental safety and operational flexibility (Ren, Li and Li, 2009). Bioleaching had a significant impact on changes in partitioning of heavy metals from mine tailings by indigenous sulfur-oxidizing bacteria (Liu et al., 2008). Similarly, Chen and Lin (2004) reported that the bioleaching method is effective when applied to remove heavy metals effectively from the livestock sludge.

4.2.7. Bioaugmentation and Biostimulation

Microorganisms can degrade a wide range of organic pollutants owing to their metabolic capacity to adapt to inhospitable environments. This makes microorganisms major players in contaminated site remediation (El Fantroussi and Agathos, 2005). Bioaugmentation is the addition of bacterial cultures to a contaminated medium and frequently used in both in situ and ex situ systems. Microorganisms having specific metabolic capability are introduced to the contaminated site in order to enhance the degradation of waste material (Garima and Singh, 2014). For years, the technique has been practiced in a number of areas including agriculture, forestry and wastewater treatment (Vogel, 1996). At sites where soil and groundwater are contaminated with chlorinated ethenes, such as tetrachloroethylene and trichloroethylene, bioaugmentation is used to ensure that the *in situ* microorganisms can completely degrade these contaminants to ethylene and chloride, which are non-toxic. Monitoring of this system is difficult (Garima and Singh, 2014). Pepper et al., (2002) indicated that both cell bioaugmentation and gene bioaugmentation can be effective in enhancing the organic degradation in contaminated soils depending on the degree of contamination and the time frame available for remediation.

Biostimulation is the addition of nutrients to a polluted site in order to encourage and improve the growth of naturally occurring chemical-degrading microorganisms (Bento et al., 2005; Mrozik, 2010). According to Hazen (2010), biostimulation is one of the most mature methods of bioremediation of hydrocarbons. As demonstrated by Abdulsalam et al. (2011), used aerobic fixed bed bioreactors to study and compare biostimulation and bioaugmentation for remediation of soil contaminated with spent motor oil. The authors indicated that biostimulation option has the highest carbon dioxide generation (6 249 mg/kg) and the control with the least (4 276 mg/kg). Although after ten weeks 50, 66 and 75% of contaminants were removed by the control, bioaugmentation and biostimulation, they suggest that biostimulation option can be used to develop a realistic treatment technology for soils contaminated with spent motor oil. A study in South Africa on compost bioremediation of oil sludge by using different manures under laboratory conditions recommended the following

1. Composting techniques is cost effective and environmental friendly for field-scales as a bioremediation process to treat sites contaminated with heavy deposit of oil sludge.
2. Composting should be combined with both bioaugmentation and biostimulation processes as a useful strategy for the degradation oil sludge. These could be used as a specific bioremediation process for practical field or full scale treatment.
3. Chemical surfactants may be added to composting process to support the biosurfactant produced by microorganisms to reduce the surface tension and increase the dissolution rates of total PAHs present in oil sludge. The application of biosurfactant and biosurfactant-producing bacteria in contaminated environments should be encouraged, as they are promising techniques in environmental biotechnologies. This is due to their potentials in biodegradability and low toxicity. Therefore, it is an important aspect regarding biological

remediation technology to use biosurfactant in a large scale bioremediation process (Ubani, 2012).

4.3. Combined Remediation Technology

4.3.1. Bioelectroremediation

Basically, the application of an electric field to a polluted site may help in the mobilization of the contaminants. That mobilization makes the contaminants available for the microorganisms. At the same time, soil bacteria appear as a colloid with a surface charge. As a result, they can be moved under the effect of the electric field. The transport of bacteria, even in small distances, may help in the interaction between the bacteria and the contaminants. Finally, the electric field can be used as a transportation mechanism to introduce into the soil the nutrients and other chemicals that may facilitate the bacterial growth and development, as well as the supply of other chemicals that can contribute to the degradation of the contaminants (Lohner et al., 2009; Wick 2009)). Tables 3 and 4 highlight the advantages and disadvantages of biological and non-biological remediation techniques.

Table 3. Advantages and disadvantages of some biological technologies used in soil remediation (Castelo-Grande et al., 2003)

Technology	Advantages	Disadvantages
Landfarming	• Relative simple design and implementation; -Short treatment times (six months to two years under optimal conditions).	• Reductions of concentration greater than 95% and concentrations lower than 0.1 ppm are difficult to achieve; -The required area is high; -Dust and vapour generation during land farming aeration may cause some air quality problems
Bioventing	• Uses readily available equipment, easy to install; - Creates minimal disturbance to the treatment site; -May not require costly off gas treatment; -Easily combinable with other technologies (e.g., air sparging, oundwater extraction).	• The high concentrations may be toxic for microorganisms; -Not applicable for certain site conditions (e.g., low soil permeability); -Sometimes requires nutrients and air injection wells; -Only treats unsaturated zones of soils, and needs other methods to treat saturated zones of soils and groundwater
Natural attenuation	• The generation of less remediation waste, and less impact on the environment; -Ease to use when combined with other technologies; -No equipment down time.	• The public may not perceive the effectiveness of the process correctly; -Site characterization can be more costly and complex; - Due to monitoring, active remediation may be more economical; -The potential exists for continued migration.
Phytoremediation.	• Is much less expensive than conventional options	• Is a technology that is seasonal; -Only applicable to low profundity.
Biosparging	• Readily available equipment; -Cost competitive; -Requires no removal, treatment, storage or discharge of groundwater.	• Some interactions among complex chemical, and physical and biological processes are not well understood; -Potential for inducing migration of constituents
Bio – Rehabilitation	• In-situ -Degradation of material dissolved in infiltrated and saturated zone; -Equipment easily available.	• The hole can be obstructed by biomass or precipitation; -Continuous monitoring and maintenance.

Table 4. Advantages and disadvantages of some non-biological technologies used in soil remediation (Castelo-Grande et al., 2003)

Technology	Advantages	Disadvantages
Vitrification (Thermal).	• Ex situ vitrification is a well-developed technology; -The mobility of contaminants is reduced/eliminated; -The vitrified mass resists leaching for geologic periods of time	• The process requires intensive energy and high temperatures up to near 2000 K; -Water in soil affects operation and increases the total costs of the process; -Off gases must be collected and treated before release; -In situ vitrification is in pilot scale development.
Incineration (Thermal)	• Contaminant toxicity, as well as volume reduction is addressed by this technology. This is especially true for organic contaminants; -Widely used and available commercially	• Metals are not destroyed and end up in the flue gases or in the ashes; -Community resistance to incineration is often present; -Certain types of soils such as clay soils or soils containing rocks may need screening.
Soil Washing (Physical-Chemical)	• Reduces the volume of contaminant, therefore, further treatment or disposal is less problematic; -Commercially available.	• Contaminant toxicity is unchanged, although volume is reduced; -Less effective when soil contains a high percentage of silt and clay; -Costs associated with the disposal of the subsequent waste streams must be considered.
Soil Vapour Extraction (physical chemical). -	• Proven performance, readily available equipment, easy to install; - Minimal disturbance to site operations; -Short treatment times (6-48 months)	• Concentration reductions greater than 90% are difficult to achieve; -Effectiveness decreases when applied to sites with low permeability; -Only treats the unsaturated zone; -May require costly treatment for atmospheric discharge of the extracted vapour.
Electrokinetic (others)	• In situ technology that has small impact on environment (soil removal is not required). -Metals are actually removed from soil unlike stabilization, which leaves the metals in the soil.	• Alkaline soils reduce the effectiveness of the process; -Requires soil moisture.

CONCLUSION

Recent advancements in technology have led to high levels of industrialization and its resultant increase in waste discharge into the environment (Okeke, 2013). Soil contamination consists of chemical degradation that leads to the partial or total loss of the soils productivity as the consequence of the accumulation of toxic substances at concentrations that

exceed the soils natural defensive capacity (Juana, 2007). Remediation is applied with aims of removing and detoxifying contaminants in the soils. Remediation of heavy metal contaminated soils is necessary to reduce the associated risks thereby availing land resource for agricultural production (Wuana, and Okieimen 2011), Single remedial technology such as chemical treatment, land filling, vitrification, electrokinetic and phytoremediation methods with their treatment cost (US$ ton ¬) of 100.500, 100.500, 75.424, 20.200 and 5.40, respectively, are quite efficiency. However, they could be applied in parallel with phytoremediation whereby plants are nanotechnologically and genetically modified to boost efficiency in the remediation of contaminated soil. The discovering of these multi-approaches establishes faster and higher removal rate of the contaminants (Karami and Shamsuddin, 2010). According to Myers (1990), biodiversity prospecting offers several opportunities of which the most important is to save as much as possible of the world's immense variety of ecosystems. Biodiversity prospecting would lead to the discovery of wild plants that could clean polluted environments of the world. The South African National Environmental Management Act, no. 62 of 2008 have stressed on the minimization of pollution and protection of the environment by industries. Despite the regulation and guidelines on how to handle and manage environmental pollution, most industries pay little attention to safe disposal of hazardous waste (Quarterly Government Gazette No. 22, 2009). In South Africa, refineries generate substantial quantity of oil sludge, a hazardous substance containing many known carcinogens (Bojes and Pope, 2007). Bioremediation is a technique that utilizes the natural biological activity of microorganisms or enzymes to transform the toxic petroleum components into less toxic or harmless metabolites (Abha and Singh, 2012). Government should provide research institute with financial and human resources to aid with soil remediation methods.

Acknowledgment

This is to express our appreciation to Dr Adeyemi Aremu for his valuable input and editorial support

References

Abdulsalam, S., Bugaje, I. M., Adefila, S. S. & Ibrahim, S. (2011). Comparison of biostimulation and bioaugmentation for remediation of soil contaminated with spent motor oil. *International Journal of Environmental Science and Technology*, *8*, 187-194.

Abha, S. & Singh, C. S. (2012). Hydrocarbon Pollution: Effects on Living Organisms, Remediation of Contaminated Environments and Effects of Heavy Metals Co-contamination on Bioremediation. In: Introduction to Enhanced Oil on Recovery (EOR) Processes and Bioremediation of Oil Contaminated Sites, Romero-Zeron, L. (Ed.). InTech Publisher, China, ISBN, 978-953-51-0629-6, 186-206.

Acar, Y. B., Gale, R. I., Alshawabken, A. N., Marks, R. E., Puppal, S., Bricka, M. & Parker, R (1995). Electrokinetic Remediation: Basic and Technology Status. *Journal of Hazardous Material, 40*, 117 –137.

Acar, Y. Yakin, B., Alshawabkeh, B. & Akram N. (1993). Principles of Electrokinetic Remediation. *Environmental Science and Technology., 27*, 2638-2647.

Akiwumi, F. A. & Butler, D. R. (2008). Mining and environmental change in Sierra Leone, West Africa: a remote sensing and hydrogeomorphological study. *Environmental monitoring and assessment, 142*, 309-318.

Al-Awadhi, N., Al-Daher, R., EI Nawawy, A. & Salba, M. T. (1996). Bioremediation of oil-contaminated soil in Kuwait. I. land farming to remediate oil-contaminated soil. *Soil and Sediment Contamination., 5*, 243-260.

Ashraf, M. A., Maah, M. J. & Yusoff, I. (2014). Soil Contamination, Risk Assessment and Remediation http://dx.doi.org/10.5772/57287 accessed 23rd February 2015.

Azeez, J. O., Hassan, O. A. & Egunjobi, P. O. (2011). Soil contamination at dumpsites: implication of soil heavy metals distribution in municipal solid waste disposal system: a case study of Abeokuta, South-western Nigeria. *Soil and Sediment Contamination, 20*, 370-386.

Bento, F. M., Camargo, F. A., Okeke, B. C. & Frankenberger, W. T. (2005). Comparative bioremediation of soils contaminated with diesel oil by natural attenuation, biostimulation and bioaugmentation. *Bioresource Technology, 96*, 1049-1055.

Bojes, H. K. & Pope, P. G. (2007). Characterization of EPA's 16 priority pollutant polycyclic aromatic hydrocarbons (PAHs) in tank bottom solids and associated contaminated soils at oil exploration and production sites in Texas. *Regulatory Toxicology and Pharmacology, 47*, 288–295.

Boopathy, R. (2000). Factors limiting bioremediation technologies. *Bioresource Technology, 74*, 63-67.

Bot, A. & Benites, J. (2005). The importance of soil organic matter: Key to drought-resistant soil and sustained food production (No. 80). *Food & Agriculture Organisation.*

Cakmak, I. & Marschner, H. (1987). Mechanism of phosphorus-induced zinc deficiency in cotton. III. Changes in physiological availability of zinc in plants. *Physiologia Plantarum, 70*, 13-20.

Cameselle, C., Gouveia, S., Akretche, D. E. & Belhadj, B. (2013). Advances in electrokinetic remediation for the removal of organic contamination in soil DOI, 10.5772/54334 INTECH.

Castelo-Grande, T. & Barbosa, D. (2003). Soil Decontamination By Supercritical Extraction, Electronic Journal of Environmental, *Agricultural and Food Chemistry, 2*, 331-336.

Chambliss, S. E., Silva, R., West, J. J., Zeinali, M. & Minjares, R. (2014). Estimating source-attributable health impacts of ambient fine particulate matter exposure: global premature mortality from surface transportation emissions in 2005. *Environmental Research Letters, 9*, 104009 doi, 10.1088/1748-9326/9/10/104009.

Dave, D. & Ghaly, A. E. (2011). Remediation technologies for marine oil spills: A critical review and comparative analysis. *American Journal of Environmental Science, 7*, 423-440.

Daverey, A. & Pakshirajan, K. (2011). Recent advances in bioremediation of contaminated soil and water using microbial surfactants. In Microbes and Microbial Technology, 207-228). *Springer New York.*

Dupont, R. R. (1993). Fundamentals of bioventing applied to fuel contaminated sites. *Environmental Progress*, *12*, 45-53.

Dushenkov, V., Kumar, P. N., Motto, H. & Raskin, I. (1995). Rhizofiltration: the use of plants to remove heavy metals from aqueous streams. *Environmental Science and Technology*, *29*, 1239-1245.

El Fantroussi, S. & Agathos, S. N. (2005). Is bioaugmentation a feasible strategy for pollutant removal and site remediation? *Current Opinion in Microbiology*, *8*, 268-275.

Environment, (2015). Scoping study on the International Year of Soils 2015. http://ec.europa.eu/environment/soil/scoping_en.htm (accessed 25th February).

EPA (1994). engineering bulletin. In Situ Vitrification treatment, United States Environmental protection agency.

Evangelou, M. W., Fellet, G., Ji, R. & Schulin, R. (2015). Phytoremediation and Biochar Application as an Amendment. *In Phytoremediation* (253-263). Springer International Publishing.

FAO (2014). Nothing dirty here: FAO kicks off International Year of Soils 2015. [ONLINE] Available at: http://www.fao.org/news/ story/en/item/ [Accessed 19 January 2015).

FAO-UNEP (1983). Guideline for the Control of Soil Degradation FAO, Rome.

Fernández-Getino, A. P., Pérez, J., Albarrán, M. D. M., Lammerding, D. M. & Walter, I. (2012). Restoration of abandoned, degraded agricultural soil using composted biosolid: influence on selected soil properties. *Arid Land Research and Management*, *26*, 200-210.

Garima, T. & Singh, S. P. (2014). Application of Bioremediation on Solid Waste Management: A Review. *Journal of Bioremediation & Biodegradation*, *5*(248), 2.doi, 10, 4172/2155- 6199.1000248

Ghrefat, H. & Yusuf, N. (2006). Assessing Mn, Fe, Cu, Zn and Cd pollution in bottom sediments of Wadi Al-Arab Dam, *Jordan. Chemosphere*, *65*, 2114–2121.

GWRTAC (1996). Phytoremediation. Groundwater Remediation Technologies Analysis Center. Publication GWRTAC-TO-96-03. 425 Sixth Avenue, Regional Enterprise Tower, Pittsburgh, PA.

Hazen, T. C. (2010). Biostimulation. *In Handbook of Hydrocarbon and Lipid Microbiology*, 4517-4530. Springer Berlin Heidelberg.

Hazen, T. C., Dubinsky, E. A., DeSantis, T. Z., Andersen, G. L., Piceno, Y. M., Singh, N. & Mason, O. U. (2010). Deep-sea oil plume enriches indigenous oil-degrading bacteria. *Science*, *330*, 204-208.

Ho, S. V., Sheridan, P. N., Athmer, C. J., Heikamp, M. A., Brackin, J. M., Weber, D. & Brodsky, P. H. (1995). Integrated in Situ Soil Remediation Technology: The Lasagna Process. *Environmental Science and Technology.*, *29*, 2528 – 2534.

Jadia, C. D. & Fulekar, M. H. (2008). Phytoremediation: the application of vermicompost to remove zinc, cadmium, copper, nickel and lead by sunflower plant. *Environmental Engineering and Management Journal*, *7*, 547-558.

Karami, A. & Shamsuddin, Z. H. (2010). Phytoremediation of heavy metals with several efficiency enhancer methods. *African Journal of Biotechnology*, *9*, 3689-3698.

Karlen, D. L., Mausbach, M. J., Doran, J. W., Cline, R. G., Harris, R. F. & Schuman, G. E. (1997). Soil quality: a concept, definition, and framework for evaluation (a guest editorial). *Soil Science Society of America Journal*, *61*, 4-10.

Katsivela, E., Moore, E. R., Maroukli, D., Strömpl, C., Pieper, D. & Kalogerakis, N. (2005). Bacterial community dynamics during in-situ bioremediation of petroleum waste sludge in land farming sites. *Biodegradation*, *16*, 169-180.

Khan, F. I., Husain, T. & Hejazi, R. (2004). An overview and analysis of site remediation technologies. *Journal of Environmental Management*, *71*, 95-122.

Kim, B. K., Baek, K., Ko, S. H. & Yang, J. W. (2011). Research and field experiences on electrokinetic remediation in South Korea. *Separation and Purification Technology*, *79*, 116-123.

King, M. M., Kinner, N. E., Deming, D. P., Simonton, J. A. & Belden, L. M. (2014). Bioventing of No. 2 Fuel Oil: Effects of air flowrate, temperature, nutrient amendment and Acclimation. *Remediation Journal*, *24*, 47-60.

Kühl, K. (2010). From Brownfields to Greenfields: A Field Guide to Phytoremediation http://urbanomnibus.net/

Kuzovkina, Y. A., Knee, M. & Quigley, M. F. (2004). Cadmium and copper uptake and translocation in five Willow (Salix L.) species, *International Journal of Phytoremediation*, *6*, 269-287.

Lal, R. (2010). Beyond Copenhagen: mitigating climate change and achieving food security through soil carbon sequestration. *Food Security*, *2*, 169-177.

Lee, M. D. & Swindoll, C. M. (1993). Bioventing for in situ remediation. *Hydrological sciences journal*, *38*, 273-282.

Lohner, S. T., Tiehm, A., Jackman, S. A. & Carter, P. (2009). Coupled electrokinetic-bioremediation: applied aspects. In: Reddy KR and Cameselle C. (Eds) 2009. Electrochemical remediation technologies for polluted soils, sediments and groundwater. Wiley. *New Jersey.USA*., 389-416.

Lyer, R. (2001). Electrokinetic remediation particulate science and technology: *An International Journal.*, *19*, 219-228.

Malina, G., Grotenhuis, J. T. C. & Rulkens, W. H. (1999). The effect of temperature on the bioventing of soil contaminated with toluene and decane. *Journal of Soil Contamination*, *8*, 455–480.

Marín, J. A., Moreno, J. L., Hernández, T. & García, C. (2006). Bioremediation by composting of heavy oil refinery sludge in semiarid conditions. *Biodegradation*, *17*, 251-261.

Mmolawa, K. B., Likuku, A. S. & Gaboutloeloe, G. K. (2011). Assessment of heavy metal pollution in soils along major roadside areas in Botswana. *African Journal of Environmental Science and Technology*, *5*, 186-196.

Molden, D. & de Fraiture, C. (2004). Investing in water for food, ecosystems and livelihoods. Blue Paper. Comprehensive assessment of water management in agriculture. International Water Management Institute, Colombo.

Mrozik, A. & Piotrowska-Seget, Z. (2010). Bioaugmentation as a strategy for cleaning up of soils contaminated with aromatic compounds. *Microbiological Research*, *165*, 363-375.

Mulligan, C. N., Yong, R. N. & Gibbs, B. F. (2001). An evaluation of technologies for the heavy metal remediation of dredged sediments. *J Hazard Mater*, *85*, 145-163.

Myers, N. (1990). The biodiversity challenge: expanded hotspots analysis. *The Environmentalist*, *10*, 243-256.

Nakayama, S. M., Ikenaka, Y., Hamada, K., Muzandu, K., Choongo, K., Teraoka, H. & Ishizuka, M. (2011). Metal and metalloid contamination in roadside soil and wild rats around a Pb–Zn mine in Kabwe, Zambia. *Environmental Pollution*, *159*, 175-181.

Nedelkoska, T. V. & Doran, P. M. (2000). Hyperaccumulation of cadmium by hairy roots of Thlaspi caerulescens. *Biotechnology and Bioengineering*, *67*, 607-615.

Nedunuri, K. V., Erickson, L. E. & Gonindaraju, R. S. (1998). Modelling the role of active biomass on fate and transport of a heavy metal in the presence of root exudates, *Journal of Hazardous Substance Research*, *1*, 1-25.

Nriagu, J. O. (1992). Toxic metal pollution in Africa. *Science of the Total Environment*, *12*,1–37.

Okeke, P. N. (2013). Enhanced electrokinetic remediation of cadmium contaminated Soil. *An International Journal of Science and Technology Bahir Dar, Ethiopia*, *2*, 16-25.

Österreicher-Cunha, P., Vargas, Jr., E. D. A., Guimarães, J. R. D., de Campos, T. M. P., Nunes, C. M. F., Costa, A. & Mano, D. M. (2004). Evaluation of bioventing on a gasoline–ethanol contaminated undisturbed residual soil. *Journal of Hazardous Materials*, *110*, 63-76.

Panichayapichet, P., Nitisoravut, S. & Simachaya, W. (2007). Spatial distribution and transport of heavy metals in soil, ponded surface water and grass in a Pb-contaminated watershed as related to land-use practices. *Environment Monitoring Assessment*, *135*, 181–193.

Pavel, V. & Gavrilescu, M. (2008). Overview of ex situ decontamination techniques for soil cleanup. *Environmental Engineering and Management Journal*, *7*, 815-834.

Pepper, I. L., Gentry, T. J., Newby, D. T., Roane, T. M. & Josephson, K. L. (2002). The role of cell bioaugmentation and gene bioaugmentation in the remediation of co-contaminated soils. *Environmental Health Perspectives*, *110* (Supple 6), 943-946

Potter, C. L., Glaser, J. A., Hermann, R. & Dosani, M. A. (1999). Remediation of contaminated East River sediment by composting technology. In Leeson A and Alleman B C (eds), Bioremediation Technologies for polycyclic aromatics hydrocarbon compounds. The fifth international in situ and on-site bioremediation symposium. San Diego, California, April 19-22, 1999. *Battelle Press*, *Columbus*, 31-36.

Prasad, M. N. V. & Freitas, H. M. (2003). Metal hyperaccumulation in plants - Biodiversity prospecting for phytoremediation technology line at http://www.ejbiotechnology.info/content/vol accessed 12/02/2015.

Qiu R., Fang, X., Tang, Y., Du, S. & Zeng, X. (2006), Zinc hyperaccumulation and uptake by Potentilla Griffithii Hook, *International Journal of Phytoremediation*, *8*, 299-310.

Reddy, K. R. & Cameselle, C. (2009). Overview of electrochemical remediation technologies. In: Reddy KR & Cameselle C. (Eds). Electrochemical remediation technologies for polluted soils, sediments and groundwater. Wiley. New Jersey. USA., 2009., 3-28.

Reid, B. J., Jones, K. C., Semple, K. T. & Fermor, T. R. (2000). Bioremediation potentials of PAHs in compost. In Leeson A and Alleman BC (eds), Bioremediation Technologies for polycyclic aromatics hydrocarbon compounds. *Battelle*, *Columbus.*, *25*-30.

Ren, W. X., Li, P. J., Geng, Y. & Li, X. J. (2009). Biological leaching of heavy metals from a contaminated soil by Aspergillus niger. *Journal of Hazardous Materials*, *167*, 164-169.

Riser-Roberts, E. (1992). Bioremediation of Petroleum Contaminated Sites, CRC Press, Boca.

Robles-González, I. V., Fava, F. & Poggi-Varaldo, H. M. (2008). A review on slurry bioreactors for bioremediation of soils and sediments. Microbial Cell Factories, Doi,10.1186/1475-2859-7-5.

Robles-González, V., Galíndez-Mayer, J., Rinderknecht-Seijas, N. & Poggi-Varaldo, H. M. (2012). Treatment of mezcal vinasses: A review. *Journal of Biotechnology*, *157*, 524-546.

Salt, D. E., Smith, R. D. & Raskin, I. (1998). Phytoremediation. *Annual Review of Plant Biology*, *49*, 643-668.

Schnoor, J. L. (1997). Phytoremediation. Technology Evaluation Report TE-98-01, Ground-Water Remediation Technologies Analysis Centre, Pittsburgh, PA. http://www.clu-in.org/download/toolkit/phyto_e.pdf.

Schnoor, J. L., Light, L. A., Mccutcheon, S. C., Wolfe, N. L. & Carreira, L. H. (1995). Phytoremediation of organic and nutrient contaminants. *Environmental Science and Technology*, *29*, 318A-323A.

Sharma, H. D. (2004). Environmental Engineering Site Remediation, Waste Contaminant, and Emerging Waste Management Technologies. Hoboken, N.J. Willey.

Smith, J., Abegaz, A., Matthews, R., Subedi, M., Orskov, B., Tumwesige, V. & Smith, P. (2014). What is the potential for biogas digesters to improve soil fertility and crop production in Sub-Saharan Africa?. *Biomass and Bioenergy*, *70*, 58-72.

Smith, P. (2008). Land use change and soil organic carbon dynamics. *Nutrient Cycling in Agroecosystems*, *81*, 169-178.

Tomei, M. C. & Daugulis, A. J. (2013). Ex situ bioremediation of contaminated soils: an overview of conventional and innovative technologies. *Critical Reviews in Environmental Science and Technology*, *43*, 2107-2139.

Treville, A. (2011). Strategic Environmental Assessment as a tool for limiting landonsumptionhttp://www.iaia.org/SpecialMeetings/prague11/proceedings/papers/Paper Treville_SEA-LandUse.pdf. (Accessed 25th February 2015).

Ubani, O. (2012). Compost bioremediation of oil sludge by using different manures under laboratory conditions. Master of Science Degree in environmental sciences at the University of South Africa supervisors Atagana HI, Thantsha MS.

Vidali, M. (2001). Bioremediation. An overview. *Pure and Applied Chemistry*, *73*, 1163-1172.

Virkutyte, J., Sillanpaa, M. & Latostenmea, P. (2002). Electrokinetic Soil Remediation Critical Overview. *The Science of the Total Environment*, *289*, 97 – 121.

Vogel, T. M. (1996). Bioaugmentation as a soil bioremediation approach. *Current Opinion in Biotechnology*, *7*, 311-316.

Wada Shin – Ichiro & Umegaki, Yuki. (2001). Major ion and electrical potential distribution in soil under electrokinetic remediation. *Environmental science and technology*, *35*, 2151 – 2155.

Wania, F. (1999). On the origin of elevated levels of persistent chemicals in the environment. *Environmental Science and Pollution Research*, *6*, 11-19.

Watanabe, M. E. (1997). Phytoremediation on the brink of commercialization. *Environmental Science and Technology*, *31*, 182-186.

Whitman, T. & Lehmann, J. (2009). Biochar—One way forward for soil carbon in offset mechanisms in Africa? *Environmental Science and Policy*, *12*, 1024 1027.

Wick, I. Y. (2009). Coupling electrokinetics to the bioremediation of organic contaminants: principles and fundamental interactions. In: Reddy KR, Cameselle C. (Eds).

electrochemical remediation technologies for polluted soils, sediments and groundwater. Wiley. *New Jersey. USA., 369*-387.

Wuana, R. A. & Okieimen, F. E. (2011). Heavy Metals in Contaminated Soils: A Review of Sources, Chemistry, Risks and Best Available Strategies for Remediation. ISRN Ecology ID 402647, 20 pages http://dx.doi.org/ "http://dx.doi.org/10.5402/2011/402647".

Yadav, B. K., Siebel, M. A. & van Bruggen, J. J. (2011). Rhizofiltration of a heavy metal (lead) containing wastewater using the wetland plant Carex pendula. *CLEAN–Soil, Air, Water, 39*, 467-474.

In: Soil Management
Editor: Manuel Esteban Lucas Borja
ISBN: 978-1-63483-274-8

Chapter 5

CAN MICROBIOLOGICAL SOIL PROPERTIES RECOVER IN THE MID-TERM AFTER WILDFIRES, CLIMATE CHANGE AFFECTION AND POST-FIRE THINNING? A CASE STUDY IN ALEPPO PINE STANDS IN THE SOUTHEASTERN OF SPAIN

***Javier Hedo*[1] *and Manuel Esteban Lucas Borja*[2]**
[1]Department of Plant Production and Agricultural Technology, School of Advanced Agricultural Engineering, Castilla La Mancha University, Albacete, Spain
[2]Department of Agroforestry Technology and Science and Genetics, School of Advanced Agricultural Engineering, Castilla La Mancha University, Albacete, Spain

ABSTRACT

Wildfires affecting forest ecosystems and post-fire silvicultural treatments may cause relevant changes in soil properties. Soil plays an important role in the fertility and stability of the forest ecosystem as it highlights the microorganisms that accomplish reactions to release soil nutrients for vegetation development. Wildfires and post-fire silvicultural treatments may cause relevant changes in forest soil properties, as well as droughts. The aim of this work is to investigate some enzyme activities in semiarid and dry Aleppo pine stands. Different plots affected by a fire 17 years ago, with or without thinning as silvicultural treatment five years after the fire, were selected. Moreover, an experiment of induced drought was carried out in half of these plots. The main finding was that soil enzyme activities did not show significant differences among burnt and unburnt plots, neither by thinned and unthinned plots, nor plots with and without induced drought experiment. We conclude that the vegetation recovery normalises post-fire soil microbial parameters after fire, as in the case of the thinning treatment. We also state that enzyme activities resist water stress, at least the 15% rainfall reduction.

Keywords: soil properties, fire, thinning, induced drought, semiarid

INTRODUCTION

Wildfires are considered natural disturbations that become part of the Mediterranean ecosystem dynamics (Naveh 1975). Moreover, they are one of the major problems in these ecosystems. Over the past few decades, number and total burnt surface of wildfires have increased in the most countries of the Mediterranean region (Moreno and Oechel 1994). Fire regime is affected by three main factors (Pausas 2004): land use changes, climate change and the increase of population. The leading cause of the increasing of the size and number of fires is the land use changes, but the climate factors have greatly contributed, due to the decreasing of the precipitations, and to the increasing of the temperatures, more acutely in the driest season, when plants show the lowest moisture level, which leads to a fuel raise by accumulation (Piñol et al., 1998; Pausas 2004; Trigo et al., 2006; Founda and Giannakopoulos 2009). Different researchers forecast that the annual burnt area average by fires get increase about 150-220% in 2090 comparing with 2000 databases (Khabarov et al., 2014). Wildfires in terrestrial ecosystems, as a direct effect of heating, remove and modify vegetation cover. Indirectly, is the soil which will transfer the impact of fire on ecosystems at the long-term, showing spatial and temporal changes in soil properties (Mataix-Solera and Cerdà 2009). Fire influences in the soil formation, as it modifies the nutrient cycles (Raison et al., 2009), physical and chemical soil properties (Úbeda and Outeiro 2009) and microbiological processes (Mataix-Solera et al., 2009).

In the Mediterranean region, the major cause of soil degradation is water erosion. Water erosion is greatly aided by the intensifying of the forest fires. Fire effects on soils are function of heating quantity (fire intensity) and the combustion time (Neary et al., 1999). Mataix-Solera and Guerrero (2007) indicate that the soil may suffer both direct and indirect changes following the occurrence of such disturbances. Direct changes are produced by direct heating of the soil during the fire and indirect changes arise from the new microclimatic situation after vegetation loss, the new ash coverage, and the rising of the water and wind erosion susceptibility (Mataix-Solera 1999).

Soil is the basis of forest ecosystems (Mataix-Solera 2007). Therefore it is important to pay attention to this system, since after its degradation can be very difficult to recover. Several studies have shown how soil microbe activities behave in arid environments (e.g., Bastida et al., 2009; García-Orenes et al., 2010; Lucas-Borja et al., 2011; Camara-Ferreira et al., 2014; Curiel-Yuste et al., 2014; Fterich et al., 2014) being few research works related to forest ecosystems (Lucas-Borja et al., 2011; Camara-Ferreira et al., 2014; Curiel-Yuste et al., 2014). Individual soil properties are not enough to assess soil quality (Nannipieri et al., 1990) as they present seasonal and spatial variations. For this reason, soil quality depends on a very large number of physical, chemical, biological, microbiological and biochemical properties (Blònska and Januszek, 2013).

Moreover, climate change is currently one of the major threats of the world ecosystems, even more for the Mediterranean forests. In addition to the fire regime modifications, the intrinsic character of the changes that climate change involves may modify also soil properties (Brevik 2013). Silvicultural treatments as thinning also disturbed soil status, though several studies state that thinning improves some soil properties such as microbial biomass or net soil CO_2 efflux, in the short term (Grady and Hart 2006), as well as the metabolic efficiency of microorganisms and soil organic matter mineralisation (Wic-Baena et al., 2013).

Diverse studies have used a combination of several variables for the purpose of make an approximation of the soil status (Dick et al., 1996). Physical and chemical properties are less used than the microbiological and biochemical properties, because they are altered only when the soil is subjected to drastic changes (Filip 2002).

Instead, microbiological and biochemical parameters are more sensitive to smooth changes (Gil-Sotres et al., 2005). Enzymatic activities have been proposed, among other variables, to evaluate soil status (Trasar-Cepeda et al., 1998).

Besides, general indicators of microbial activity have been largely used in forest soil status characterization (Armas et al., 2007), as basal soil respiration and microbial biomass carbon.

MATERIAL AND METHODS

Study Area

Two large fires affected 14,000 and 30,000 hectares in Calasparra and Yeste in the provinces of Murcia and Albacete, respectively, both located in the southeastern of the Iberian Peninsula. The fire occurred in summer 1994 and affected mature *Pinus halepensis* Mill. stands. The potential vegetation at the first site corresponds to the *Rhamno lycioidis-Quercetum cocciferae* S., while Yeste corresponds to the *Bupleuro Rigidi-Querceto rotundifoliae* S. (Rivas-Martínez 1987). The two study sites are on the mesomediterranean floor, which is semiarid at Calasparra and dry at Yeste. The average temperature and rainfall for the study area are, respectively, 282 mm Calasparra and 16.3ºC, and 503 mm and 13.5ºC for Yeste (AEMET). Both sites presented high post-fire natural regeneration, with an average density of about 7,000 trees/ha in Yeste and of 45,000 trees/ha in Calasparra. According to the Spanish Soil Map (Guerra Delgado 1968), Yeste and Calasparra soils are classified as Inceptisols and Aridisols, respectively. Soil texture at both sites is classified as loam/clay-loam.

Materials and Experimental Design

In December 2011, three plots (150 m^2, 10m x 15m) in the burnt area of Yeste and three in the burnt area of Calasparra were selected. Some others were selected in the unburnt areas surrounding the perimeter of each fire. In all five subplots, the plots were established randomly as 4 m^2 (2m x 2m), in which six composite samples and distributed random subsamples were taken to make the samples as representative as possible. After removing litter, samples were sieved through a 2-mm sieve and were homogenised in the laboratory.

In the burnt plots and in order to evaluate the effects of thinning (5 years after the fire event) and drought (3-year rainfall reduction) on soil properties, two levels of rainfall (natural rainfall vs. rainfall reduction or induced drought) and two levels of thinning (presence and absence of thinning) were combined in four treatments with three replicates. In July 2009, two experimental sites, naturally burnt in summer 1994, were selected: Yeste and Calasparra. Twelve plots (10 m x 15 m) were set up inside each experimental site (six thinned plots and six unthinned plots). Thinned and unthinned plots were occupied by high Aleppo pine post-fire natural regeneration (45,000 saplings/ha in Calasparra, 7,000 saplings/ha in Yeste).

Thinning operations were carried out in 1999 with up to 1,600 saplings/ha at both the Calasparra and Yeste sites. Each group of six plots, as described before, was split into two groups at each experimental site as follows: three plots receiving natural rainfall and the other three plots occupied by PVC gutters to simulate drought-induced conditions, occupying about 15% of the plot surface (Cotillas et al., 2009).

This percentage claimed to intercept around 15% of precipitation according to the IPCC (2007) predictions for the Mediterranean Basin in next years.

Microbiological Soil Properties

There have been used three enzyme activities, each one related to one of the nutrient cycles (nitrogen, phosphorus and carbon cycle). Urease activity (UA) was determined as the NH_4^+ released in the hydrolysis reaction (Kandeler et al., 1999).

Alkaline phosphatase (PA) and β-glucosidase (BA) activities were measured following the methods reported by Tabatabai and Bremner (1969) and Tabatabai (1982), respectively.

Results

Urease, β-glucosidase and phosphatase were not affected by any of the studied variables. Fire, thinning and the induced drought showed not significant differences between its levels (Table 1).

Differences in urease and beta-glucosidase values are shown between Yeste and Calasparra.

Yeste showed higher values in beta-glucosidase while Calasparra showed higher values in urease (Figures 1, 2 and 3). In terms of phosphatase values, both Yeste and Calasparra showed similar values.

Discussion

The results have showed that measured soil properties do not differ between burnt and mature stands, but differ from Yeste (dry climate) and Calasparra (semiarid climate).

The predominant vegetation in both study areas is Aleppo pine. We can state that the differences observed between Calasparra and Yeste were not caused by vegetation.

In relation to thinning, no significant differences were found in the behaviour of these three enzymatic activities. Aleppo pine is a serotinous species that exhibits good post-fire natural regeneration, showing good post-fire seedling recruitment (Leone et al., 2000).

Table 1. Results of the two-factor ANOVA (site and fire) for microbiological properties and microbiological ratios

	Urease		Phosphatase		Beta-glucosidase	
	F ratio	*p-value*	F ratio	*p-value*	F ratio	*p-value*
Factors						
Fire	45.15	0.9246	0.37	0.5486	5.58	0.8134
Thinning	0.01	0.9932	0.29	0.754	1.7	0.1993
Induced drought	0.02	0.9819	0.05	0.9519	0.72	0.4948

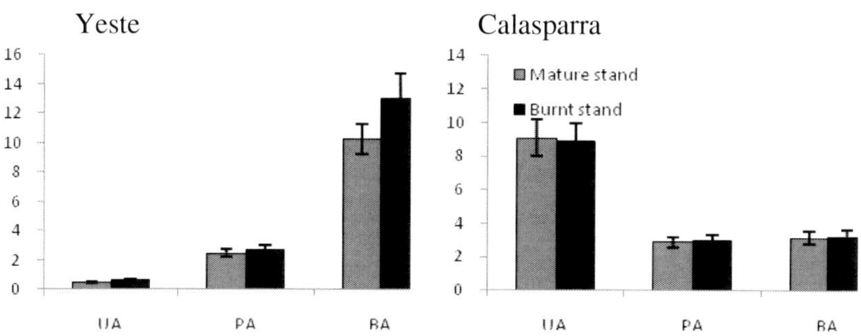

Figure 1. Urease activity (μmol N-NH_4^+ g^{-1} dry soil $hour^{-1}$), phosphatase activity (μmoles PNP g^{-1} dry soil $hour^{-1}$) and beta-glucosidase activity (μmoles PNP g^{-1} dry soil $hour^{-1}$) for the studied mature and burnt stands. The error bars shown are the LSD intervals at $p<0.05$.

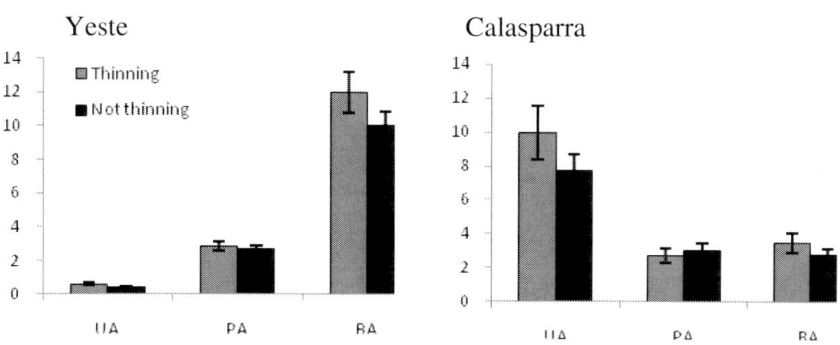

Figure 2. Urease activity (μmol N-NH_4^+ g^{-1} dry soil $hour^{-1}$), phosphatase activity (μmoles PNP g^{-1} dry soil $hour^{-1}$) and beta-glucosidase activity (μmoles PNP g^{-1} dry soil $hour^{-1}$) for the thinned and not thinned plots. The error bars shown are the LSD intervals at $p<0.05$.

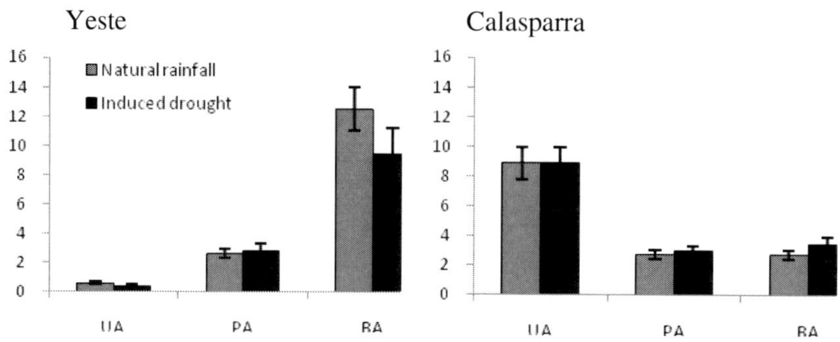

Figure 3. Urease activity (µmol N-NH_4^+ g^{-1} dry soil $hour^{-1}$), phosphatase activity (µmoles PNP g^{-1} dry soil $hour^{-1}$) and beta-glucosidase activity (µmoles PNP g^{-1} dry soil $hour^{-1}$) for the plots without and with the induced rainfall. The error bars shown are the LSD intervals at $p<0.05$.

In this context, temporary plant cover loss and subsequent plant recruitment after a fire event may enhance the microbiological soil properties recovery. According to our results, enzymatic activities capacity recovery should be achieved 17 years after the fire and after the thinning.

Related to the induced drought, no significant differences were also found in the studied enzymatic activities. Aponte et al., (2010) stated that drought affects Mediterranean soil microbial dynamics. In our study, the induced drought did not affect the three studied enzymatic activities. Thus, microbial communities in dry and semiarid regions might show some stability and water stress resistance strategy when faced with drought events, in our case the predicted rainfall reduction by the IPCC (2007).

References

Aponte C., Marañón T., García L. V. 2010. Microbial C, N and P in soils of Mediterranean oak forests: influence of season, canopy cover and soil depth. *Biogeochemistry,* 101, 77–92.

Armas C. M., Santana B., Mora J. L., Notario J. S., Arbelo C. D., Rodríguez-Rodríguez A. 2007. A biological quality index for volcanic Andisols and Aridisols (Canary Islands, Spain): variations related to the ecosystem development. *Science of the Total Environment*, 378, 238–244.

Bastida F., Pérez-de-Mora A., Babic K., Hai B., Hernández T., García C., Scholer M. 2009. Role of amendments on N cycling in Mediterranean abandoned semiarid soils. *Applied Soil Ecology*, 41, 195-205.

Blońska E., Januszek K. 2013. Usability of enzyme activity in estimation of forest soil quality. *Folia Forestalia Polonica*, 55 (1), 18–26. DOI: 10.2478/ffp-2013-0003.

Brevik E. C. 2013. The Potential Impact of Climate Change on Soil Properties and Processes and Corresponding Influence on Food Security. *Agriculture,* 3, 398-417. DOI:10.3390/agriculture3030398.

Câmara Ferreira A. C., Carvalho Leite L. F., Ferreira de Araújo A. S., Eisenhauer N. 2014. Land-use type effects on soil organic carbon and microbial properties in a semi-arid region of northeast Brazil. *Land Degradation and Development*, DOI: 10.1002/ldr.2282.

Curiel-Yuste J., Fernandez-Gonzalez A. J., Fernandez-Lopez M., Ogaya R., Penuelas J., Sardans J., Lloret F. 2014. Strong functional stability of soil microbial communities under semiarid Mediterranean conditions and subjected to long-term shifts in baseline precipitation. *Soil Biology and Biochemistry,* 69, 223-233.

Cotillas M., Sabaté S., Gracia C., Espelta J. M. 2009. Growth response of mixed Mediterranean oak coppices to rainfall reduction. Could selective thinning have any influence on it? *Forest Ecology and Management*, 258, 1677–1683.

Dick R. P., Breakwell D. P., Turco R. F. 1996. Soil enzyme activities and biodiversity measurements as integrative microbiological indicators. In Methods for assessing soil quality. Doran J. W., Jones A. J. (eds) 247-271.

Filip Z. 2002. International approach to assessing soil quality by ecologically-related biological parameters. *Agriculture, Ecosystems and Environment,* 88, 169–174.

Fterich A., Mahdhi M., Mars M. 2014. The effects of Acacia tortilis subsp. Radianna, soil texture and soil depth on soil microbial and biochemical characteristics in arid zones of Tunisia. *Land Degradation and Development,* 25, 143-152. DOI: 10.1002/ldr.1154.

Founda D., Giannakopoulos C. 2009. The exceptionally hot summer of 2007 in Athens, Greece — A typical summer in the future climate? *Global and Planetary Change,* 67(3–4):227–236.

García-Orenes F., Guerrero C., Roldán A., Mataix-Solera J., Cerdà A., Campoy M., Zornoza R., Bárcenas G., Caravaca, F. 2010. Soil microbial biomass and activity under different agricultural management systems in a semiarid Mediterranean agroecosystem. *Soil and Tillage Research*, 109 (2), 110-115. DOI: 10.1016/j.still.2010.05.005.

Gil-Sotres F., Trasar-Cepeda C., Leirós M., Seoane S. 2005. Different approaches to evaluating soil quality using biochemical properties. *Soil Biology and Biochemistry,* 37, 877-887.

Grady K. C., Hart S. C. 2006. Influences of thinning, prescribed burning, and wildfire on soil processes and properties in South-western ponderosa pine forests: a retrospective study. *Forest Ecology and Management*, 234, 123–135.

Guerra Delgado A. CSIC. 1968. Leyenda del Mapa de Suelos de España (1:1.000.000). Inst. Nac. Edafol. CSIC, Madrid.

IPCC 2007. Climate Change 2007: Impacts, Adaptation and Vulnerability. Contribution of Working Group II to the Fourth Assessment Report of the Intergovernmental Panel on Climate Change ML, Parry O. F., Canziani J. P., Palutikof P. J., van der Linden, C. E. Hanson (eds.), Cambridge University Press, Cambridge, UK, p. 976.

Kandeler E., Stemmer M., Klimanek E. 1999. Response of soil microbial biomass, urease and xylanase within particle size fractions to long-term soil management. *Soil Biology and Biochemistry*, 31, 261-273.

Khabarov N., Krasovskii A., Obersteiner M., Swart R., Dosio A., San-Miguel-Ayanz J., Durrant T., Camia A., Migliavacca M. 2014. Forest fires and adaptation options in Europe. *Regional Environmental Change*, DOI: 10.1007/s10113-014-0621-0.

Lucas-Borja M. E., Candel D., Jindo K., Moreno J. L., Andrés M., Bastida F. 2011. Soil microbial community structure and activity in monospecific and mixed forest stands, under Mediterranean humid conditions. *Plant and Soil,* 354, 359–370.

Mataix-Solera J., Cerdà A. 2009. Incendios forestales en España. Ecosistemas terrestres y suelos. En: Cerdà, A. y Mataix-Solera (eds). Efectos de los incendios forestales sobre los suelos en España. El estado de la cuestión visto por los científicos españoles. Càtedra de Divulgació de la Ciència. Universitat de Valencia, pp: 25-53.

Mataix-Solera J., Guerrero C., García-Orenes F., Bárcenas G. M., Torres M. P. 2009. Forest fire effects on soil microbiology. En: Cerdà A and Robichaud PR (Eds.). Fire effects on soil and restoration strategies. Sciences Publishers, pp 225-256.

Mataix-Solera J., Guerrero C. 2007. Efectos de los incendios forestales en las propiedades edáficas. En: Mataix-Solera J. (ed.), Incendios Forestales, Suelos y Erosión Hídrica. Caja Mediterráneo CEMACAM Font Roja-Alcoi, Alicante, pp 5-40.

Mataix-Solera J. 1999. PhD thesis: Alteraciones físicas, químicas y biológicas en suelos afectados por incendios forestales. Contribución a su conservación y regeneración. Facultad de Ciencias. Universidad de Alicante. España.

Moreno J. M., Oechel W. C. (eds.). 1994. The Role of Fire in Mediterranean-Type Ecosystems. Springer-Verlag, Ecological Studies 107, New York.

Naveh Z. 1975. The evolutionary significance of fire in the Mediterranean region. *Vegetatio*, 29, 199-208.

Nannipieri P., Grego S., Ceccanti B. 1990. Ecological significance of the biological activity in soil. In: Bollag J. M., Stotzky G. (eds.) *Soil Biochemistry*, vol 6. Marcel Dekker: New York, 293–355.

Neary, D. G., C. C. Klopatek, L. F. DeBano, Ffolliott P. F. 1999. Fire effects on belowground sustainability: a review and synthesis. *Forest Ecology and Management,* 122: 51-71.

Pausas J. G. 2004. La recurrencia de incendios en el monte mediterráneo. In: Vallejo R. (ed.). Avances en el estudio de la gestión del monte mediterráneo. pp. 47-64. CEAM, València.

Piñol J., Terradas J., Lloret F. 1998. Climate Warming, Wildfire Hazard, and Wildfire Occurrence in Coastal Eastern Spain. *Climatic Change*, 38, 345–357.

Pausas J. G. 2004. Changes in fire and climate in the eastern Iberian Peninsula (Mediterranean basin). *Climatic Change*, 63, 337–350.

Rivas Martínez S. 1987. Memoria del mapa de series de vegetación de España (1:400.000). ICONA. Ministerio de Agricultura, Pesca y Alimentación, Madrid.

Raison R. J., Khanna P. K., Jacobsen K. L. S., Romanya J., Serrasolses I. 2009. Effect of fire on forest nutrient cycles. En: Cerdà A, Robichaud PR (Eds.). Fire effects on soil and restoration strategies. *Sciences Publishers*, pp 225-256.

Trasar-Cepeda C., Leirós C., Gil-Sotres F., Seoane S. 1998. Towards a biochemical quality index for soils: an expression relating several biological and biochemical properties. *Biology and Fertility of Soils,* 26, 100–106.

Tabatabai M. A., Bremner J. M. 1969. Use of p-nitrophenyl phosphate for assay of soil phosphatase activity. *Soil Biology and Biochemistry*, 1: 301-307.

Tabatai M. A. 1982. Soil enzymes. In: Methods of analysis, part 2, 2nd ed. Agronomy Page et al. (eds.) 9, 389-396.

Trigo R. M., Pereira J. M. C., Pereira M. G., Mota B., Calado T. J., Dacamara C. C., Santo F. E. (2006) Atmospheric conditions associated with the exceptional fire season of 2003 in Portugal. *International Journal of Climatology,* 26 (13):1741–1757.

Wic-Baena C., Andrés-Abellán M., Lucas-Borja M. E., Martínez-García E., García-Morote F. A., Rubio E., López-Serrano F. R. 2013. Thinning and recovery effects on soil properties in two sites of a Mediterranean forest, in Cuenca Mountain (South-eastern of Spain) *Forest Ecology and Management,* 308, 223-230.

In: Soil Management
Editor: Manuel Esteban Lucas Borja

ISBN: 978-1-63483-274-8

Chapter 6

BIOCHAR ADSORPTION PROPERTIES AND THE IMPACT ON NAPHTHALENE AS A MODEL ENVIRONMENTAL CONTAMINANT AND MICROBIAL COMMUNITY DYNAMICS: A TRIANGULAR PERSPECTIVE

N. Maurathan, C. H. Orr and T. K. Ralebitso-Senior*
Teesside University, Middlesbrough, UK

ABSTRACT

It has been suggested that biochar can sorb contaminants and influence soil microbial communities. Since this is a relatively new research area, literature on triangulating biochar, contaminants and microbial population dynamics is limited. Nevertheless, seminal studies (2008-2015) that entailed different biochars – feedstock and pyrolysis conditions – have been used to highlight the challenges that must be recognised while addressing fundamental knowledge gaps on contaminant sorption by biochar, and how this interaction affects bacterial community phenotypic, phylogenetic, functional and compositional structure. For simplicity, and based on existing literature, naphthalene has been used as a model contaminant. Thus factors that potentially influence its sorption in general, and by biochar in particular, are considered. Furthermore, some determinants of microbial response are discussed since emergent studies have shown that specific bacterial taxa respond differently to different biochars, including increase and/or decrease in abundance and changes in biomass and community composition. Naturally, the well-established responses of bacteria to naphthalene are analysed within the context of the potential for biochar-facilitated (bio)remediation. Generally, a link between molecule (naphthalene) degradability by bacteria in the presence of biochar remains a considerable knowledge gap. This highlights the essential need for comprehensive cross-/multi-disciplinary and multi-phasic research that should underpin directed and sustainable application of biochar for contemporary and historical contaminant amelioration.

* Corresponding author: T. K. Ralebitso-Senior. School of Science and Engineering, Teesside University, Middlesbrough, TS1 3BA, UK. E-mail: K.Ralebitso-Senior@tees.ac.uk.

Keywords: biochar, environmental contaminants, naphthalene, sorption, microbial ecology dynamics

INTRODUCTION

Complex bacterial communities are found in natural soil ecosystems where they show greater diversity exceeding that of eukaryotic organisms (Torsvik and Øvreås 2002). The compositions of these communities are dynamic and respond faster than eukaryotes to natural and anthropogenic environmental changes (Valverde et al., 2011). The growing interest in the application and exploitation of biochar encapsulate its, as yet, unrealised potential to address the ever growing challenges of climate change mitigation, contaminant attenuation and the global demand for increased agricultural productivity. Although the interactions between biochar and soil microbial populations are recognised, a considerable paucity of knowledge exists of microbial diversity and functional responses to biochar addition. Its distinctive physical, chemical and biological properties have, however, highlighted tremendous exploitation potential in key sustainable biotechnologies, which are underpinned by complex interacting microbial communities. Consequently, studies have shown that biochar can influence soil microbial communities (Anderson et al., 2011) and determine the fate of contaminants due to its sorption capabilities (Beesley et al., 2010). If the material is to be used sustainably and successfully, particularly for enhanced bioremediation, then an in-depth understanding of its dual impact on these two essential components is critical.

BIOCHAR DEFINITION

Biochar is a carbon rich charred product, usually produced from biomass such as: agricultural and forestry (e.g., logging, wood processing) residues; herbaceous crops; aquatic and marine biomass; and wastes including municipal, industrial, livestock and poultry wastes. It is derived commonly through pyrolysis under oxygen limited conditions mainly for subsequent soil application as an amendment or conditioner, and to sequester carbon and enhance soil fertility (Lehmann and Joseph 2009; Warnock et al., 2010; Duku et al., 2011; Anderson et al., 2011; Matovic 2011; Peng et al., 2011; Ohsowski et al., 2012; Panwar et al., 2012).

During pyrolysis, feedstock biomass undergoes three parallel pathways namely: char and gas formation; liquid and tar formation; and gasification and carbonization (Shafizadeh 1982 cited in Lehmann and Joseph, 2009). Cai et al., (2011) simplified these pathways into gas, liquid and solid phases, respectively, depending on the pyrolysis conditions.

Generally, biomass lignocellulosic composition and proportion vary with different substrates and, thus, influence the thermal degradation/pyrolysis product characteristics (Duku et al., 2011), which include density, particle size, volatile matter, and contents of ash, carbon and moisture. These are affected subsequently by changes in the biomass constituents during pyrolysis, for example, cross-linking, depolymerisation, fragmentation (Duku et al., 2011), attrition, cracking and microstructural rearrangements (Verheijen et al., 2010).

Parameters such as process temperature, heating and volatile substance removal rate, and process or reaction residence time can influence these pathways further and, ultimately, the final product properties. Thus high volatilization, carbonization and gas formation result in products with high carbon and low oxygen, nitrogen, hydrogen and sulphur content at temperatures >600°C (Lehmann and Joseph, 2009).

Bacterial Response to Biochar Application

Several researchers have reported that parallel to the effects of its physico-chemical characteristics, the concentrations and ratios of bioavailable C and N on biochar surfaces can affect microbial community biomass, structure, composition, dominance and functional activity (Nguyen and Lehmann, 2009; Steinbeiss et al., 2009; Hilscher and Knicker, 2011; Jindo et al., 2012). Thus the need to understand soil bacterial community response to biochar has been recognised.

The underpinning research has consisted of the application of several methods such as phospholipid fatty acid analysis (Steinbeiss et al., 2009; Santos et al., 2012), leucine analysis (Jones et al., 2012), polymerase chain reaction-based denaturing gradient gel electrophoresis (PCR/DGGE) (Feng et al., 2012), automated ribosomal intergenic spacer analysis (ARISA) (Khodadad et al., 2011), CO_2 efflux (Steinbeiss et al., 2009), terminal restriction fragment length polymorphism (TRFLP) (Anderson et al., 2011) and pyrosequencing (Anderson et al., 2011; Feng et al., 2012) to measure changes in biomass, abundance, diversity, community composition and structure and shifts in functional capacities.

Shifts in Abundance

For changes in abundance, an unweathered biochar produced from Monterey pine (*Pinus radiate*), with sintered glass as the control, was used by Anderson et al., (2011) to assess its influence on bulk and rhizosphere soil microbial communities in the presence of ryegrass (*Lolium* sp.). The PCR/ TRFLP-based analyses revealed increased relative abundances for different families and genera in the biochar augmented bulk and rhizosphere soils between weeks 2 and 12 of the study (Table 1).

These included members belonging to the Bradyrhizobiaceae and Hyphomicrobiaceae families, which are involved in nitrogen cycling and, hence, supported suggestions that biochar can provide a transient soil 'nitrogen island'. Also, the pine-derived biochar promoted the proliferation of *Mycobacterium* species, which have been reported to degrade polyaromatic hydrocarbons following release from biochar (Hennessee et al., 2009).

Notwithstanding, Anderson et al. (2011) recorded a decrease in the relative abundance of some bacterial families in response to unweathered *Pinus radiate*-derived biochar supplementation. Thus a decline in abundance in bulk and rhizosphere soil communities resulted for Streptomycetaceae (*Streptomyces* spp), Micromonosporaceae, Conexibacteraceae (*Conexibacter* spp), Micrococcaceae (*Arthrobacter* spp), Micromonosporaceae and Nocardioidaceae (*Marmoricola* spp).

Table 1. Different bacterial families and genera that showed relative increases and decreases in abundance in bulk soil, the rhizosphere or both during a 12-week study by Anderson et al. (2011)

Family	Genera	Bulk soil	Rhizosphere
Increased abundance			
Acidothermaceae		*	*
Bradyrhizobiaceae	*Rhodoblastus Rhodopseudomonas Bradyrhizobium Nitrobacter*	*	
Cellulomonadaceae	*Cellulomonas*	*	*
Rhizobiaceae/Geodermatophilaceae	*Geodermatophilis*	*	*
Frankiaceae		*	*
Hyphomicrobiaceae	*Rhodoplanes Starkeya*	*	
Leptospiraceae/Sphingomonadaceae		*	
Microbacteriaceae	*Microbacterium Agroccoccus*	*	*
Micromonosporaceae			*
Mycobacteriaceae	*Mycobacterium*	*	*
Nitrosomonadaceae	*Nitrosovibrio*		*
Nocardioidaceae			*
Rhodospirillaceae		*	*
Streptosporangiaceae	*Streptosporangium*	*	*
Thermomonosporaceae	*Actinomadura Actinoallomurus*	*	*
Xanthomonadaceae			*
Decreased abundance			
Acidothermaceae			*
Bradyrhizobiaceae			*
Burkholderiales/Micrococcaceae/ Xanthobacteraceae		*	*
Conexibacteraceae/Cystobacterineae	*Conexibacter*	*	*
Leptospiraceae/Sphingomonadaceae			*
Micromonosporaceae		*	
Nitrosomonadaceae	*Marmoricola*	*	
Nocardioidaceae		*	
Streptomycetaceae	*Streptomyces sp.*	*	*
Xanthomonadaceae	*Xanthomonas*	*	

Changes in Community Structure

As for abundance, biochar addition can effect positive, negative or neutral changes in bacterial community structure (Table 2). For example, a 12-week study by Jindo et al., (2012) recorded an increase in bacterial biomass due to supplementation with poultry manure mixed with hardwood biochar compared to manure only, with a specific increase in the ratio of Gram-positive to Gram-negative bacteria. Also, the authors observed that biochar addition resulted in a considerable decrease in bulk density of the composting pile and, thus, increased aeration, probably accounted for enhanced bacterial colonisation.

A 3-year study by Jones et al., (2012), using agricultural soils planted with maize and grass, revealed increased bacterial growth and biomass but a reduced ratio between fungi and bacteria during the first two years. This trend changed in the third year where similar fungal to bacterial biomass ratios were recorded, including for the control soil without the *Fraxinus excelsior*- and *Fagus sylvatica*-derived biochars.

Table 2. The main responses by bacterial taxa and/or strains to biochars produced from different feedstock and processes. The biochars were tested individually or mixed with manures in specific soil systems

Feedstock	Process	Tested system	Key responses	References
Pine-*Pinus radiate*		Ryegrass (*Lolium perenne*) planted soil-bulk soil and rhizosophere	Increase in abundance e.g., *Bradyrhizobiaceae*, *Hyphomicrobiaceae*, *Streptosporangineae* and *Thermomonosporaceae*	(Anderson et al., 2011)
			Decrease in abundance e.g., *Streptomycetaceae*, *Micromonosporaceae*, *Bradyrhizobiaceae* and *Hyphomicrobiaceae*	
Corn stalk	Biochar reactor - pyrolysis	Paddy soils: Inceptisol and Ultisol	Increase in abundance of methanotrophic proteobacteria	(Feng et al., 2012)
Hardwood tree	Japanese traditional kiln -pyrolysis	Poultry and cow manure	Increased Gram-positive to Gram-negative bacteria ratio	(Jindo et al., 2012)
Chipped trunks and large branches of *Fraxinus excelsior* and *Fagus sylvatica*	Pyrolysis	Agriculture soil planted with maize and grass	Bacterial dominated community shift	(Jones et al., 2012)
Oak heartwood (Quercus laurifolia Michx.); Eastern gama grass (Tripsacum dactyloides)	Custom-built pyrolyzer - pyrolysis	Agricultural field planted with maize and grass	Abundance of bacteria belonging to the Actinobacteria and Gemmatimonadetes phyla	(Khodadad et al., 2011)
Pine wood (*Pinus ponderosa*)	Combustion	Forest subsoil and granite soil	Pyrogenic organic matter utilized preferably by Gram-positive bacteria; No changes in the microbial community composition	(Santos et al., 2012)
Glucose (GB); Yeast (YB)	Hydrothermal pyrolysis	Arable soil and forest soil	GB utilized preferably by bacteria	(Steinbeiss et al., 2009)

Negative impacts of biochar on microbial biomass have been reported as exemplified by Steinbeiss et al., (2009) who applied hydrothermal yeast- and glucose-derived biochars on arable and forest soils. Yeast biochar decreased the biomass of Gram-positive and Gram-negative bacteria by 7-14% in the two soils during a 4-week study. Also, statistically significant reductions were recorded for total bacterial and fungal biomass in response to the addition of the 0% N glucose biochar.

Furthermore, soil organic carbon concentration measurements and CO_2 evolution analysis revealed that yeast-based biochar addition enhanced soil microbial activity/metabolism considerably in both soils during the first 12 weeks of the study. Degradation of the yeast-derived material was largely effected by the soil fungi than bacteria hence their biomass increased and decreased, respectively.

In contrast, although an increased utilization of the glucose-based biochar was recorded for bacteria than for fungi, little or no increase in microbial activity was recorded in its presence (Steinbeiss et al., 2009). Therefore, the authors concluded that the different effects were probably due to condensation (high condensed glucose-derived and low condensed yeast-derived biochar), nitrogen content and chemical structures of the chars.

Several studies have also reported no shifts in microbial community structure due to biochar addition. Santos et al., (2012) conducted a 6-month study where they applied pine wood (*Pinus ponderosa*) combusted at 450°C to forest subsoil and granite soil.

Although the pyrogenic matter was utilized preferably by Gram-positive bacteria, no changes were recorded in the community composition of this group compared to Gram-negatives, fungi and Actinobacteria. Also, similar compositions resulted in the char-augmented and control soils between days 42 and 180.

Neutral effects have also been recorded depending on feedstock and/or addition ratio. Thus, Jindo et al., (2012) recorded no significant difference in bacterial biomass for hardwood augmentation of cow manure while Steinbeiss et al., (2009) observed similar total microbial biomass profiles for arable and forest soils supplemented with 5% N yeast-derived biochar. In the same study, addition of glucose-based biochar did not make any bacterial compositional changes in either soil.

Parallel to investigations on contemporary biochar application, researchers have studied shifts in microbial community diversity, richness, abundance and structure in historical biochar-augmented soils, including with particular focus on their metabolic/growth characteristics. Thus although analyses of *Terra preta* cannot provide an exact profile of its microbial community compositional/structural genesis (O'Neill et al., 2009; Zimmerman et al., 2011; Glaser and Birk, 2012), investigations have revealed increased abundance of k-selected bacteria such as actinomycetes while r-selected communities were predominant in adjacent reference soils (O'Neill et al., 2009). The researchers suggested that k-strategists have slower carbon degradation and/or mineralisation rates and a decreased capacity to adapt to new carbon sources. Consequently, this metabolic feature has also been proposed to explain the high recalcitrance of *Terra preta.* In contrast, r-selected strategists are characterised by higher adaptation to new carbon sources and increased carbon mineralisation capacity.

MICROBIAL RESPONSE TO NAPHTHALENE CONTAMINATION

Naphthalene Contamination

Polycyclic aromatic hydrocarbons (PAHs) are a group of chemicals that contain two or more fused aromatic rings and are among the most common and persistent environmental contaminants (Seo et al., 2009). Since some PAHs are considered probable human carcinogens, their presence in the environment is of concern (Menzie et al., 1992). Naphthalene ($C_{10}H_8$), an anthropogenic PAH with two fused benzene rings, has been reported as one of the priority polyaromatic organic pollutants by the US Environmental Protection Agency and is one of the most commonly found contaminants (Pathak et al., 2009; Patel et al., 2012). Formed by degradation of crude oil, tars and coke and produced as an intermediate or commodity chemical, it is the simplest and most abundant PAH found in many industry-impacted sites (Bayard et al., 2000; Purwaningsih et al., 2004; Ania et al., 2007; Dsikowitzky et al., 2011). Also, naphthalene is a non-polar hydrophobic organic compound that occurs in a solid state at room temperature and has a water solubility of 25-30 mgL^{-1} at ambient temperature. As a result, it is typically present in the aqueous phase in contrast to high molecular weight PAHs that are often present in non-aqueous solid phases of soils and sediments. According to Alves et al. (2008), naphthalene poses a significant risk to human health due to its concomitant toxic, carcinogenic and mutagenic properties. Its high concentrations, persistence and mobility in the environment have raised considerable concern (Pathak et al., 2009) hence soil ecosystems with naphthalene concentrations $\geq$120 $mgkg^{-1}$ require remediation (Li et al., 2011).

Biodegradation and Bioavailability

The capacity to degrade organic contaminants is widespread among soil bacteria as many have metabolic pathways that facilitate utilization of the molecules as carbon and/or nitrogen sources (Puglisi et al., 2007; Di Gennaro et al., 2009; Haritash and Kaushik, 2009; Lu et al., 2011; Mao et al., 2012). Thus indigenous microbial communities often adapt to contaminant exposure by increasing in number and producing catabolic enzymes. Also, historical exposure to compounds with similar molecular structures can induce the adaptation of existing metabolic pathways to degrade new contaminants (Singer et al., 2003).

According to Lucas et al., (2008), the biodegradation of polymers is a complex process with several, often simultaneous, steps including biodeterioration, biofragmentation, assimilation and mineralisation. Biodeterioration and biofragmentation are the breakdown of heavy molecular weight macromolecules into low molecular weight subunits often as a result of catabolic enzyme secretion by specific organisms. The subunits can then enter the cytoplasm or remain in the extracellular environment. Cellular assimilation occurs where low molecular weight molecules are metabolised to produce energy and secondary metabolites. Mineralisation allows complete oxidation of these metabolites under aerobic or anaerobic conditions.

Bioavailability is the concentration of a substrate available for microbial degradation (Barret et al., 2010). Therefore, the dissolved fraction of hydrophobic organic substances (Sobisch et al., 2000), and substrates that are sorbed to particles, under specific circumstances (Barret et al., 2010), are considered bioavailable. For example, the composition of soil organic matter (SOM) (i.e., high and low content natural organic matter) influences the bioavailability of organic contaminants. Also, covalent bonding (oxidative coupling) of contaminants to soil humic acids results in increased recalcitrance compared to contaminants sorbed to soil surfaces (Burgos 1999). Subsequently, a sorptive phase that is more available than an aqueous phase results in reduced biodegradability but, potentially, enhanced remediation. Notwithstanding, enhanced contaminant availability can result in increased concentration and, thus, biotoxicity including to the potentially catabolic strains (Sobisch et al., 2000). Generally, bioavailability depends on the mass transfer rate, which is determined by the diffusion coefficient (concentration gradient, concentration at source, concentration at the sink/cell, diffusion path length and aqueous equilibrium affected by surface area and porosity of SOM); production of extracellular polymeric substances (EPS) for biofilm formation; and release of bioemulsifiers/ biosurfactants (Mittal and Rockne 2009; Pathak et al., 2009). These factors are, in turn, affected by microbial population density and metabolic activity (Johnsen et al., 2005; Barret et al., 2010; Chen et al., 2011; Megharaj et al., 2011). Furthermore, bacterial biosurfactants which can interact with biofilms (Schreiberová et al., 2012), are affected by changes in pH and salinity (Abouseoud et al., 2010) and can, therefore, effect different bacterial responses.

Due to decreased water solubility and high soil-water partitioning, microbial degradation of some PAHs is limited with biodegradation dependent on: the physico-chemical properties of contaminants; soil characteristics; environmental parameters; sorption/desorption coefficients and contaminant bioavailability; and the composition, structure and catabolic capacity of the indigenous microbial community. In addition, the hydrophobicity of high-molecular weight PAHs, in particular, allows interactions with soil organic matter which make them more recalcitrant to microbial catabolism. This is exacerbated by the typical co-occurrence of PAHs at contaminated sites with homologous heterocyclic compounds that are similarly resistant to microbial degradation (Sobisch et al., 2000; Johnsen et al., 2005; Haritash and Kaushik 2009; Mittal and Rockne 2009; Barret et al., 2010; Alberti et al., 2012; Ma et al., 2012; Mao et al., 2012).

According to Li and Chen (2009), surfactants, including bio-derived moieties produced by some soil microorganisms, can increase PAH bioavailability by partitioning from the solid phase into the aqueous phase and, hence, facilitate biodegradation. However, surfactant sorption into the soil matrix reduces their mobility, increases their adsorption affinity to PAHs and, as a result, reduces the biodegradable PAH in the aqueous phase. Moreover, surfactant composition or molecular structure affects the aqueous and sorptive phases of PAHs (Sobisch et al., 2000). For example, naphthalene adsorption by alkylbenzenesulfonate, an anionic surfactant, decreased with increasing aqueous surfactant concentration (0 - 1000 mgL^{-1}) while naphthalene sorption by the cationic cetylrimethyl ammonium bromide increased with the aqueous phase concentration up to 100 mgL^{-1} (Zhang et al., 2009). Differences in the sorption affinity of these surfactants were explained by the availability of their hydrophobic adsorption sites and subsequent adsorption to soil.

Naphthalene Biodegradation Pathways

As for other PAHs, naphthalene can be metabolised via aerobic and anaerobic pathways with most bacteria mineralising it under aerobic conditions (Xu-Xiang et al., 2006). However, different naphthalene degrading strains can use different metabolic pathways, with an example proposed in Figure 1 (Seo et al., 2009), hence the benzoate and β-ketoadipate pathways were used simultaneously by *Nocardia otitidiscaviarum* strain TSH1 (Zeinali et al., 2008).

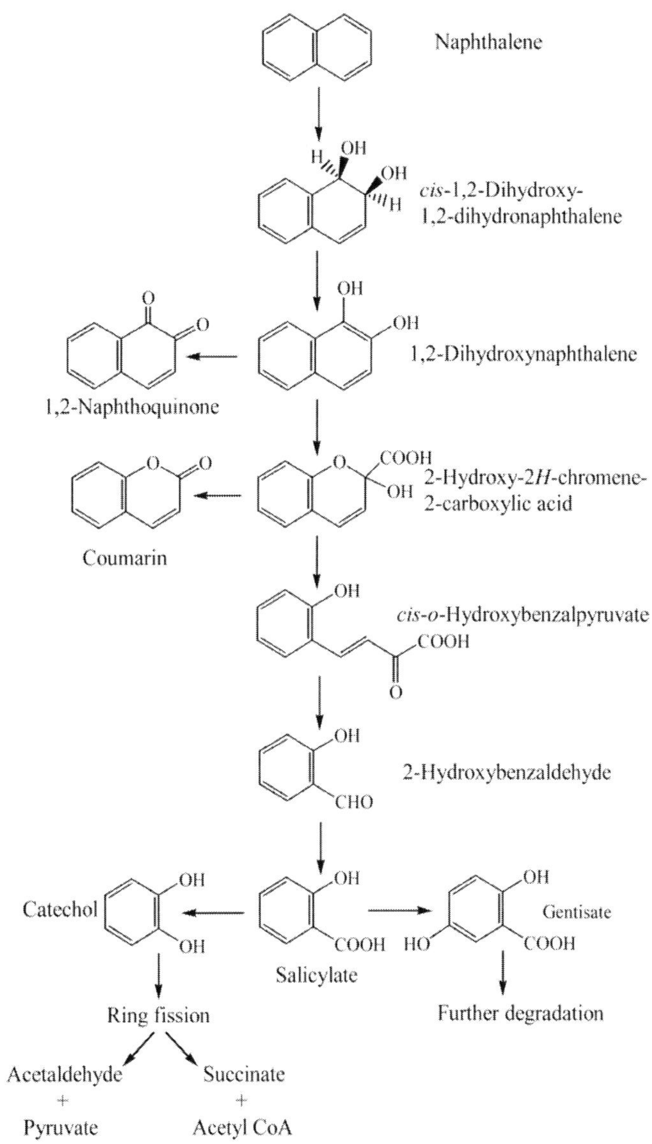

Adapted from Seo et al., 2009.

Figure 1. A proposed naphthalene biodegradation pathway.

Typically, the initial step in naphthalene catabolism is the production of cis-naphthalene dihydrodiol (cis-1,2-dihydroxy-1,2-dihydronaphthalene) catalysed by naphthalene dioxygenase. Several intermediates are produced subsequently until catechol in aerobic pathways, which is mineralised finally to water and carbon dioxide (Xu-Xiang et al., 2006).

Numerous enzymes are involved in the naphthalene degradation pathways and are encoded by multiple genes including *nahAa*, *nahAb*, *nahAc*, *nahAd*, *nahB*, *nahF*, *nahC*, *nahQ*, *nahE* and *nahD*, with most located within the nahABCDEF operon. These encode naphthalene conversion to salicylate while the nahGHIJK operon is catalytic for the production of pyruvate and acetaldehyde from salicylate (Yen and Gunsalus, 1985; Mrozik et al., 2003; Ono et al., 2007).

Naphthalene Contamination, Catabolic Bacteria and Community Structure

Measuring bacterial community response to different and changing environmental conditions is complex and challenging. This is exacerbated for naphthalene degrading microbial communities, which are demonstrably distinct from the total PAH degrading community, and their compositional changes relative to differences in contaminated sites (Muckian et al., 2007). Nevertheless, since it is the simplest and most soluble, naphthalene is used often as a model PAH to investigate the degradative capacity of bacterial strains and communities (Mrozik et al., 2003).

Although the ever-expanding list of microorganisms that utilize naphthalene as a carbon source includes *Alcaligenes, Burkholderia, Mycobacterium, Polaromonas, Ralstonia, Rhodococcus, Sphingomonas*, and *Streptomyces* species, it is dominated by the *Pseudomonas* genus (Seo et al., 2009), hence its representative mesophilic Gram-negative strains have been used extensively to elucidate naphthalene/ PAH catabolism (Zeinali et al., 2008). Singleton et al., (2005) used stable isotope probing and identified naphthalene degrading bacteria with a predominance of members belonging to the *Pseudomonas*, *Ralstonia* and *Acidovorax* genera. Also, 359 entries were recorded in the NCBI database (Accessed 23/02/15) for the *nahAc* gene, which codes for the initial dioxygenase enzyme in the naphthalene biodegradation pathway.

Screening and cultivation are the primary steps used conventionally to identify and categorise PAH degrading organisms on the basis of cell morphology, proteins and cell wall fatty acids. Consequently, several approaches have been used to isolate and identify (novel) indigenous naphthalene catabolising bacteria. For example, Kafilzadeh et al., (2011) used petroleum oil-contaminated soil and culture-based techniques to isolate *Bacillus, Pseudomonas*, *Corynebacterium*, *Staphylococcus* and *Micrococcus* species, which could degrade 86%, 80%, 77%, 69% and 58% naphthalene, respectively, in 7 days. Also, Patel et al., (2012) isolated three degradative Proteobacteria, *Achromobacter* sp., *Pseudomonas* sp. and *Enterococcus* sp. from a petroleum shipyard site.

Since, *ca* 1-5% of microorganisms in any environment are culturable, culture-based techniques may obviate complete characterisation of naphthalene degraders that are as yet uncultivated. Culture-based analysis suggested that Sphingomonadaceae and Actinobacteria are efficient PAH degraders however, experimental accounts often show different dominant taxa (Martin et al., 2012). As a result, several molecular techniques including PCR-based DGGE and TRFLP, biosensors, DNA hybridization and sequencing techniques have been applied to study PAH degrading microbial community composition, structure and phylogenetic relationships (Shi et al., 2010). For example, Uhlik et al., (2012) used pyrosequencing to identify abundant naphthalene degraders and again recorded a predominance of Pseudomonads but also unculturable *Rhodanobacter*, *Burkholderia*, *Pandorea* and *Dyella* spp. Other unique naphthalene catabolic bacteria that have been identified with ecogenomic tools include the halotolerant Ochrobactrum sp. VA1 (Arulazhagan and Vasudevan 2011), denitrifying γ-Proteobacteria and Actinobacteria (Lu et al., 2011) and thermophilic Nocardia and Bacillus species (Zeinali et al., 2008).

Total and indigenous naphthalene degrading communities show structural and compositional changes in response to naphthalene concentration with shifts in the dominance of different genera reported, for example, by Gomes et al., (2005). Specifically, the researchers used 16S rDNA analysis and recorded dominances of *Burkholderia* sp. RP007 and *Nocardiavinaceae* species in naphthalene contaminated microcosm soil. This composition changed when the contaminated soil was inoculated with a genetically modified (GM) *Pseudomonas putida* strain containing a pNF142 plasmid. The presence of the GM strain decreased the complexity of the indigenous *β*-Proteobacteria and Actinobacteria communities reducing the number of bacteria capable of degrading naphthalene (Gomes et al., 2005).

Martin et al., (2012) spiked microcosms with different concentrations of labelled PAHs, used qPCR and sequence analysis and observed shifts in the microbial community with increases in novel catabolic Betaproteobacteria such as *Acidovorz*, *Rhodoferax*, *Hydrogenophaga* and *Polaromonas*. Ji et al., (2013) suggested that the shifts could remain beyond contaminant degradation.

Responses of Naphthalene Degrading Bacteria

Naphthalene degrading bacteria, like any microbial community, respond to changes in environmental parameters such as soil moisture content, soil amendments (Filonov et al., 1999), nutrient availability (e.g., N, P), substrate concentration, pH, temperature, salinity, aerobic and anoxic conditions (Pathak et al., 2009; Lin et al., 2010; Lu et al., 2011). Most commonly tested parameters for bacterial response and naphthalene biodegradability include pH, temperature, inoculum size, growth rate and molecule concentration.

Optimal conditions for its degradation vary between microorganisms hence *Pseudomonas* sp. HOB1 isolated from sediments showed maximum naphthalene degradation at pH 7.5-8.5, with 2,000 ppm degraded within 24 hours, and maximum tolerance of 60,000 ppm naphthalene at an optimal temperature range of 35-37°C (Pathak et al., 2009). *Bacillus fusiformis* performed best at 30°C and neutral pH and degraded 99% of 50 mgL^{-1} naphthalene within 96 hours (Lin et al., 2010). *Streptomyces* sp. ERI-CPDA-1 showed an ability to degrade >99% of naphthalene at an initial concentration of 0.1% (v/v) in neutral pH, 1-5 gL^{-1} NaCl salinity and 30°C within 7 days (Balachandran et al., 2012). Patel et al., (2012) investigated community-based degradability using a model association consisting of *Achromobacter* sp. BAB239, *Pseudomonas* sp. DV-AL2, *Enterobacter* sp. BAB240 and *Pseudomonas* sp. BAB241 and observed complete degradation of 1,000 ppm naphthalene within 36 hours at 37°C and pH 8-10. Also, they observed that substrates such as peptone, yeast extract, sodium succinate and sodium acetate, as well as increased oxygen availability, enhanced the biodegradability.

Changes in bacterial growth rates are considered as responses to naphthalene. Negligible lag phase followed by an exponential phase which lasted until complete catabolism was observed for *Pseudomonas sp.* HOB1 strain by Pathak et al., (2009). The initial degradation lag phase was attributed to a decreased cell number (inoculum size), which when increased exponentially, resulted in rapid degradation in subsequent stages of growth. Also, the degradative biomass remained high probably due to the metabolism of naphthalene intermediates (Pathak et al., 2009). During the exponential growth phase, *Pseudomonas putida* showed increased adhesion to naphthalene contaminated soil.

According to Hwang et al. (2009), the hydrophobicity of bacterial cell walls associated with naphthalene assimilation during logarithmic growth enhanced soil affinity. Furthermore, nutrient addition can increase bacterial biomass, change the growth rate and affect naphthalene degradability (Lee et al., 2003). Thus a study by Filonov et al., (1999) revealed that bacterial counts doubled following naphthalene supplementation of a soil model system containing macro and micro nutrients.

Additional bacterial responses to naphthalene include changes in cell membrane structure and integrity due to the compositional and structural alterations of cell-derived fatty acids. The latter is attributed to the possible induction or inhibition of enzymes that catalyse saturated and unsaturated fatty acids synthesis (Mrozik et al., 2004; Mrozik et al., 2005). For example, naphthalene degrading *Pseudomonas putida* and *Pseudomonas stutzeri* synthesised new fatty acids following exposure to the molecule (Mrozik et al., 2005).

Also, chemotactic responses have been reported for PAH degrading bacteria in general. Chemoreceptors play a key role in chemotactic responses and signalling, particularly within motile bacteria hence some *E. coli* species have five chemoreceptors. *Pseudomonas putida* strains showed chemotactic and aerotactic, i.e., energy taxis, responses to PAH (Krell et al., 2012). These facilitate bacterial migration towards more favourable oxygen tension and redox/oxygen environments (Nichols and Hardwood, 2000). Another response of naphthalene degrading bacteria is biofilm (EPS) formation, one of key mechanisms for bacterial strains/communities to evade environmental stress. Thus, the naphthalene-catabolizing *Pseudomonas stutzeri* formed thick ring-shaped biofilms under laboratory conditions (Shimada et al., 2012).

Bioremediation Potential

Naphthalene is comparatively more biodegradable than high molecular weight PAHs hence bacteria can adapt rapidly following exposure to the contaminant (Puglisi et al., 2007; Di Gennaro et al., 2009; Mao et al., 2012). Therefore, there is potential to ameliorate naphthalene contaminated ecosystems using indigenous microorganisms, including cultivable microbial strains that can use naphthalene as a carbon and energy source (Mrozik et al., 2003; Pawar et al., 2013). Land remediation can, however, be difficult since soil, a heterogeneous medium, is not necessarily characterised by conditions that are conducive/optimum for naphthalene catabolism. Also, as already discussed, its hydrophobicity can result in binding of organic matter and reduced bioavailability (Mrozik et al., 2003; Johnsen et al., 2005). A study using multi-contaminated wasteland revealed that although PAHs and a catabolic microbial community were detected, little degradation occurred due to reduced molecule bioavailability (Cébron et al., 2013). In contrast, spiking the soil with 'fresh' phenanthrene showed biodegradation concomitant with increases in the number of degradative microorganisms.

Surfactants such as SDS, TritonX-102 and Brij 35 can increase the bioavailability of hydrophobic compounds (Mrozik et al., 2003). Although few organisms produce surfactants in the environment, their production by naphthalene degrading bacteria has been considered a response, which enhances bioavailability by increasing the surface area of the hydrophobic molecule and, thus, increases its biodegradability. Biosurfactants can be of low molecular weight such as rhamnolipids, trehalose lipids and sophorolipids or high molecular weight biopolymers (Ron and Rosenberg, 2002). They can be classified further as endo and exo where the former accumulate on cell surfaces and the latter occur in the surrounding environment such as the culture medium. For example, analogues of caprylic acid were identified as exo-surfactants in naphthalene utilization by *Pseudomonas putida* and *Burkholderia* sp. (Puntus et al., 2005). *Pseudomonas sp.* HOB1 strain produced biosurfactant but addition of external surfactant did not promote the catabolic potential of the strain (Pathak et al., 2009). Similarly, addition of Tween 80, sodium dodecyl sulfate (SDS) and cetyl trimethyl ammonium bromide (CTAB) did not produce a statistically significant effect on the biodegradability of naphthalene by a consortium of *Achromobacter* sp. BAB239, *Pseudomonas* sp. DV-AL2, *Enterobacter* sp. BAB240 and *Pseudomonas* sp. BAB241 (Patel et al., 2012).

Many fungi are capable of oxidising PAHs using cytochrome P-450 monoxygenase in a process similar to the transformation of lignocellulosic aromatic compounds (Johnsen et al., 2005). Therefore, Cébron et al., (2013) proposed inoculum supplementation with organic matter containing fungal mycelia to increase the bioremediation potential of naphthalene degrading bacteria. The researchers spiked soil with PAHs and recorded an increase in the ratio of fungi to bacteria. Specifically, saprophytic fungi such as Fusarium solani, capable of degrading high molecular weight organic compounds, became dominant with a subsequent increase in Gram-negative Pseudomonas species.

Some bacterial strains have been modified genetically as potential bioremediation inoculants. Bioluminescent *Pseudomonas fluorescens* HK44 was inoculated into soils with and without PAH contamination and the indigenous microbial community dynamics observed over a 14-year period (Ji et al., 2013). Although the PAH contamination was ameliorated completely following inoculation, the GM bacterium could not be recovered from the soil. Also, pyrosequencing showed a long-term shift in the microbial community structure in the contaminated soil. This was in contrast to the work by Gomes et al., (2005) who recorded a dominance of an inoculated GM *Pseudomoas putida*, which outcompeted the indigenous PAH degrading community in contaminated microcosms.

BIOCHAR AS SORBENT

Detailed discourse on sorption is outwith the scope of this chapter so the reader is referred to other comprehensive reviews (e.g., Alberti et al., 2012). Briefly, the authors highlighted that the physico-chemical properties of a sorbent, and environmental factors such as temperature and pH, influence its sorption behaviour while surface area, porosity, solubility, concentration, hydrophobicity, aromacity, functional groups, polarity, van der Walls forces, π-interactions and ionic properties are a few of many factors that can be used to explain its specific sorption mechanisms. Thus, sorption mechanisms (e.g., ionic exchange, reverse phase), kinetics and thermodynamic sorption models, which consider several parameters have, in turn, been used to describe the overall sorption phenomena. Examples of sorption models that use two parameters include the Langmuir, Freundlich, Jovanovic and Temkin isotherms whereas examples of models that consider three or more parameters include the Redlich–Peterson, Sips (or Langmuir–Freundlich), Tóth, Unilan isotherm, Dubinin–Radushkevich, Radke–Prausnitz, Fritz–Schlünder and Extended Langmuir isotherms. As a result, choosing a suitable model is crucial to predicting the dynamic sorption process of any material.

In parallel to the applicability of biochar for enhanced soil fertility and greenhouse gas sequestration, the material has been recognised as a functionally and economically viable sorbent (Beesley et al., 2010; Xu et al., 2012). Thus it can immobilise organic contaminants such as polycyclic aromatic hydrocarbons (Oleszczuk et al., 2012), herbicides (Nag et al., 2011), inorganics (Wang et al., 2015), pesticides (Zheng et al., 2010), (poly) chlorinated compounds (Xu et al., 2012), phosphates (Yao et al., 2011) and metals (Inyang et al., 2012; Ruthiraan et al., 2015; Yan et al., 2015).

Sorption Mechanisms

i. Adsorption

This entails the transfer of molecules (solutes) onto the surface particles of a sorbent or into its bulk in solid phase from a liquid (Alberti et al., 2012). It is relatively a non-linear competitive process and adsorption of organic contaminants is influenced by the carbonaceous sorbents available in soil. For biochar, increasing pyrolytic temperature produces more carbonised organic matter than amorphous matter, increases surface area and, thus, favours non-linear adsorption of organic contaminants. It can be proposed, therefore, that biochar surface area is the principal determinant of its adsorption properties.

However, due to competitive inhibition or blocking of sorption sites by co-existing organic compounds, the adsorption affinity for specific organic contaminants can be weakened for aged biochar, in particular, where the sorption sites are usually highly saturated (Beesley et al., 2011).

ii. Partitioning

Linear, non-competitive absorption (i.e., partitioning sorption mechanism) by non-carbonised organic matter in biochar is influenced largely by nano- and meso-porosity, which are dependent on pore width, volume and filling (Beesley et al., 2011). Low pyrolysis temperature chars favour partitioning sorption. For example, atrazine partitioning into organic phase due to relatively high organic carbon content was suggested as the reason for 77% sorption by dairy manure-derived biochar produced at 200° C (Cao and Harris, 2010). In contrast, absorption of hexachlorobenzene by wheat straw-derived char produced at ≥500°C was explained by linearity absorption pattern (Song et al., 2012).

iii. Other Factors

In addition to porosity and surface area, factors that influence the sorption properties of biochar include functional groups, π-interactions, surface coverage, contribution percentage, electrical multilayer adsorption, condensation in capillary pores (Lou et al., 2011), hydrogen bonding (Qui et al., 2009) and pore-filling (Oleszczuk et al., 2012). For example, sorption of pentachlorophenol, a hydrophobic organic toxic compound, was affected by the surface carboxyl and lactone fuctional groups of rice straw biochar (Lou et al., 2011). Also, phenolic hydroxyl groups, associated π-interactions and intermolecular hydrogen bonding interactions were proposed by Qui et al., (2009) as some of the key sorption mechanisms for organic dyes.

Sorption-Based Interactions Between Biochar and Naphthalene

Although an emergent research topic with increasing literature on biochar sorption of different contaminants (Ahmad et al., 2014), its potential application for naphthalene is still relatively limited. As a result, studies published between 2008 and 2014 on naphthalene sorption by biochars in (soil) ecosystems are summarised in Table 3. As already discussed, the physico-chemical and functional properties of biochars vary depending on the type of biomass, process and parameters. Indeed, Chen and Chen (2009) reported that the sorption functional properties of biochar can differ even when the same feedstock has been used.

Therefore, specific biochar characteristics are considered relative to the respective experimental design/objectives with different biochar-specific mechanisms for naphthalene sorption explored individually.

For analysis of sorption properties with naphthalene as the model molecule, Chen and Chen (2009) conducted a batch study where orange peel-derived biochars were pyrolyzed under oxygen limited conditions, a residence time of six hours and a temperature range of 150-700°C. All products were sieved (0.154 mm) and naphthalene sorption isotherms determined in calcium chloride with biodegradation suppressed by NaN_3. Ten different concentration point isotherms were prepared and equilibrated with rotation (20 rpm) for 7 days at room temperature in the dark.

The soft and hard carbon surface functional groups of the biochar were analysed using FTIR spectra. Aromaticity (H/C), polar molar ratios {(O + N)/ C} and O/C were calculated, and the feedstock decomposition temperatures for pectin, cellulose and hemicelluloses determined.

Biomass loss was recorded with increasing pyrolysis temperature and was consistent with the biochar yield at 150-400°C. Thus, thermogravimetric and derivative thermogravimetric measurements of biomass loss and composition relative to pyrolysis temperature showed decreases due to loss of water, pectin, hemicelluloses and cellulose at specific temperatures.

Specifically, biochar FTIR spectra showed that –OH groups started to disappear and disappeared completely at temperatures >200°C and >400°C, respectively, whereas C-O-C groups disappeared completely above 300°C. Also, CH_2 and ester C = O groups disappeared from biochar surfaces at temperatures >350°C. Only the aromatic C = C and C = O were preserved in pyrolysis temperatures above 500°C. Generally, "soft" carbon components of biomass were eliminated but the aromatic "hard" carbons were recorded with increasing pyrolysis temperature (Chen and Chen 2009).

Table 3. Summary of biochar-specific effects on naphthalene as determined by biochar feedstock, process conditions and characteristics

Biochar substrate	Pyrolysis temperature	Biochar characteristics	Effect on naphthalene	References
Pine needles	100 – 700°C		Sorbant aromaticity increases with increasing pyrolytic temperature. Biochars produced at low temperatures remove napthalene via partitioning while biochars produced at high temperatures remove napthalene via adsorption.	(Chen et al., 2008)
Orange peel	150 - 700°C.		Biochar produced at 700°C showed higher sorption affinity to naphthalene due to high surface area, high aromaticity and porosity.	(Chen and Chen 2009)
Orange peel treated with ferric and ferrous chlorides	250°C, 400°C 700°C.	Magnetic	Magnetic biochar produced at 400°C showed higher sorption affinity.	(Chen et al., 2011)
Maize stover and paper mill waste	600°C		Paper mill waste derived biochar showed higher sorption affinity due to high aromaticity and high surface area. Feedstock composition and contact time of biochar and naphthalene also influenced the sorption behaviour.	(Oleszczuk et al., 2012)
Biochar substrate	Pyrolysis temperature	Biochar characteristics	Effect on naphthalene	References
Waste wood pellets	520°C		Biochar was shown to sorp phenanthrene and napthalene contamination from urban storm water runoff with 100% and 76% removal efficiency, respectively.	(Reddy et al., 2014)
Nitrogen-rich biomass	170, 250, 350 and 450°C		Normalized sorption coefficient increased with increased charring temperatures due to the reduction in O-containing polar moieties. N-rich biochars could sorp phenanthrene, naphthalene and 1-naphthol predominantly controlled by the hydrophobic interactions between the contaminants and the aromatic components in biochars.	(Zhang et al., 2014)

Calculations for biochar aromaticity (H/C), polar molar ratios {(O + N)/C} and O/C revealed decreased H/C ratio with increased pyrolysis temperature and indicated a reduced preservation of the original biomass components such as lignin, cellulose and hemicelluloses. Also, the biochars became less hydrophilic as indicated by decreased O/C ratio while a decrease in the {(O + N)/C} ratio was indicative of reduced polar functional groups. Comparisons in another study that used the same orange peel biomass (Chen et al., 2011) suggested an increase in biochar surface area with a rise in pyrolysis temperature due to the elimination of aliphatic alkyl and ester C = O groups and the appearance of aromatic lignin core (Chen and Chen, 2009). Changes in the surface area were, however, not consistent with increasing temperature. Therefore, the researchers concluded, that due to their heterogeneity, the sorption affinity of biochar, even when produced from the same feedstock, should be corroborated empirically.

These investigators also considered the influence of pyrolysis temperature on biochar sorption characteristics. Thus, high polar surface of the 150°C-biochar showed increased affinity for polar water molecules and, as a result, the adsorption sites were blocked competitively against naphthalene adsorption. The highest adsorption was recorded for 700°C-biochar due to its increased surface area and reduced polarity. Also, the naphthalene sorption isotherms showed a linear relationship with the 150°C-biochar mainly due to partition mechanisms. Nevertheless, the partition phase of molecule sorption to biochars produced at high temperatures (>300°C) was less favoured due to the weakened aliphatic phase at high temperature whereas the hard aromatic carbon phase was enhanced. At 700°C, the adsorption process was facilitated further by π-interactions between naphthalene moieties and highly aromatic biochar surfaces (Chen and Chen, 2009).

Similar to their earlier study, Chen et al., (2011) used orange peel biomass to produce non-magnetic and ferric/ferrous chloride-based magnetic biochars (0.154 mm) at different pyrolysis temperatures (250°C, 400°C and 700°C). They recorded similar naphthalene sorption isotherms except that the isotherm equilibrium was achieved in approximately four days (Chen and Chen, 2009). Also, for relationships between physico-chemical properties and magnetism, Chen et al., (2011) analysed the magnetic and non-magnetic biochars for elemental composition, surface area and pore volume. Specifically, surface functional groups such as –OH, CH_2, ester C = O, aromatic C = C, aromatic C = O, aromatic CO- and phenolic –OH were visualised with FTIR spectroscopy. Feedstock components such as lignin and some polysaccharides were preserved in magnetic biochars produced at 250 and 400°C but eliminated at 700°C. In addition, many surface functional groups, which were undetected in the 400°C non-magnetic biochars, were recorded in their magnetic equivalents. Similarly, comparatively higher H/C ratios were observed in the magnetic biochars following pyrolysis at 250 and 400°C. The 700°C magnetic biochars had decreased carbon contents that were indicative of increased iron oxide concentrations, hence their characteristic crystalline and amorphous structures. The non-magnetic biochar produced at 700°C had higher surface area and pore volume than their magnetic equivalents. These properties were reduced further in the magnetic and non-magnetic biochars produced at 400°C (Chen et al., 2011). Overall, increased naphthalene sorption resulted for the 400°C magnetic biochar compared to the 400°C non-magnetic and 250/700°C magnetic equivalents.

To test the sorption properties of biochar with naphthalene and sewage sludge as a model contaminant and waste stream, respectively, Oleszczuk et al., (2012) used slow pyrolysis at 600°C and a residence time of 20 minutes to produce biochars from maize stover (MSB) and paper mill waste (PMW). Mixtures of sewage sludge (2 g) and the two biochars (2 mm) were prepared at ratios of 0.5 - 10% (w/w) and incubated in the dark at room temperature for 30 days. The NaN_3-sterilized sludge + biochar mixtures were then equilibrated for six weeks prior to cleaning, extraction and GC analysis. Also, the influence of biochar aging on the sorption of pore water from sewage sludge PAHs, including naphthalene, were analysed during 7, 30 and 60 days.

The total (extractable) naphthalene concentration was <0.2 $mgkg^{-1}$ sewage sludge while the dissolved concentration ranged from 2 to 3 ngL^{-1}. Sorption of freely dissolved naphthalene was higher for the paper mill waste biochar at all ratios than for the maize stover product with the lowest MSB ratio (0.5% w/w) recording no sorption affinity for naphthalene.

Although both biochars showed low sorption affinity for naphthalene than other PAHs in the sewage sludge, a clear relationship was observed between enhanced sorption affinity for freely dissolved naphthalene with increasing PMW ratio. A similar trend was recorded when the contact time with the biochars was increased from 7 to 30 days.

The three steps that are involved in the mass transfer of organic contaminants from soil, sediment and sewage sludge to biochar sorption sites entail contaminant desorption from the original matrices, diffusion through water and sequestration by biochar (Oleszczuk et al., 2012). Therefore, the researchers suggested that weak pollutant binding to sewage sludge, in contrast to soil or sediment, potentially facilitated the transfer rate. In addition, since a faster PAH transfer rate from sewage to activated carbon than to biochars was recorded relative to sorption and contact time, the authors proposed that pore-filling mechanisms, where biochar pores were potentially blocked by sewage sludge organic matter, inhibited biochar-based sorption.

In summary, this study suggested that elemental composition, surface properties, feedstock composition and pyrolysis conditions determined the sorption characteristics of the biochars. Specifically, PMW showed more sorption affinity than MSB probably due to higher surface area and higher aromaticity associated with higher carbonisation.

POTENTIAL MECHANISMS FOR BIOCHAR-MICROORGANISMS-NAPHTHALENE DYNAMICS

To link biochar, microbial communities and naphthalene, the effects of biochar on specific soil parameters and the resultant responses of the microbial communities are discussed first with biochar-induced changes on naphthalene bioavailability and microbial responses considered subsequently. Overall, bacteria show different responses to biochar addition with: increased, decreased and unchanged abundance; increased and decreased biomass; and changed and consistent community structure. Researchers have proposed several mechanisms for these effects including microbial utilization of biochar-based nutrients, which differ according to the feedstock and production conditions. According to Zimmerman et al., (2011), aliphatic biochars are more susceptible to mineralisation. Thus, Steinbeiss et al., (2009) produced glucose- and yeast-derived biochars, which were highly aromatic and aliphatic, respectively. Although both were characterised by similar phenolic, carbonyl and hydroxyl polar functional groups, the highly condensed glucose-based biochar was utilized preferentially by bacteria while the nitrogen containing low condensation yeast biochar was used by fungi.

Also, biochars produced from some feedstocks such as manure and crops are nutrient rich compared to their wood-based equivalents, hence Sohi et al., (2010) suggested that nutrient rich biochars would be degraded and assimilated preferentially by environmental microorganisms. Generally, non-stable carbon constituents of biochar are readily degradable and likely to impact on soil microbial activities (Lehmann et al., 2011; Masek et al., 2011).

Thus, Anderson et al., (2011) suggested that the recorded preferential utilization of *Pinus radiata* biochar, with an increased abundance of specific bacterial groups such as *Stakeya* spp, was due to labile biochar carbon (e.g., acetate). Also, labile carbon can be utilized by soil microorganisms within the first few weeks and months of application (Masek et al., 2011).

Smith et al., (2010) stated that carbon (3.4 mg-Cg^{-1}) from young switchgrass (*Panicum virgatum*)-derived biochar (500°C) was available for microbial mineralisation shortly after application to soil.

The effects of process temperature on biochar mineralisation have been explored widely. Hilscher and Knicker (2011) recorded microbial degradation and mobilisation of 350°C-grass pyrogenic organic matter in a 24-month laboratory study. Also, Santos et al., (2012) conducted a 6-month incubation study with wood-derived pyrogenic organic matter (450°C) and observed its preferential use by Gram-positive bacteria.

According to a study by Khodadad et al. (2011), bacterial metabolic activities increased in soils amended with a 250°C-biochar compared to its 650°C equivalent produced from oak, possibly due to increased availability of labile carbon for the former. Notwithstanding, a decrease in bacterial counts was recorded for the unburned soil amended with 250°C low temperature biochar while increased numbers and diversity resulted from augmentation of burned and unburned soils with the 650°C-biochar. The researchers adopted Zimmerman's thesis (2010, as cited in Zimmerman, 2011) and proposed several mechanisms including increased surface area and porosity due to high pyrolysis temperature (650°C), which provided new suitable niches/habitats conducive to dynamic microbial colonization and metabolic activities. Although occupying the biochar micropores would protect them from larger predators, Warnock et al., (2007, cited in Gomez-Eyles et al., 2011) suggested that pore-residing microorganisms may not be metabolically active.

Mineralisation of biochar in soil can be influenced also by the process method. Thus in a 65-day soil incubation study, Bruun et al., (2011) recorded a 5.5% labile carbon loss from wheat straw biochar produced using fast pyrolysis compared to 2.9% for the equivalent slow pyrolysis product. Also, increased utilization and incorporation of the fast pyrolysis biochar-derived carbon into microbial biomass resulted during a 24-day incubation study with a similar trend for labile N. Nevertheless, biochar mineralisation does not necessarily mean assimilation of the char-derived carbon into microbial biomass. For example, isotope labelling showed only 2.5% microbial biochar-C assimilation during a 624-day study (Kuzyakov et al., 2009 cited in Zavalloni et al., 2011) while Zavalloni et al., (2011) reported no incorporation after 20 days of incubation. Therefore, microbial responses would also depend on the fraction of bioavailable and assimilable biochar carbon.

It is reasonable to expect changes in microbial community composition and structure due to biochar stability (Lehmann et al., 2011). Generally, stable biochar carbon is not readily utilized by microorganisms with its bioavailability reduced further when sorbed to soil organic matter (Lehmann et al., 2011; Masek et al., 2011). Nonetheless, biochar can be degraded by soil biotic and abiotic physico-chemical processes, which also determine its stability (Masek et al., 2011) and bioavailability. Thus, enhanced mineralization of corn-derived biochar (>600°C) and increased carbon loss were reported in the presence of water and under saturated conditions, respectively (Nguyen and Lehmann, 2009).

Changes in soil microbial community structure can also be due to the 'priming effect' of biochar on other nutrient sources such as soil organic matter. Positive priming is accelerated SOM mineralisation with subsequent enhanced soil structure aeration and moisture content, and increased labile carbon, nitrogen, microbial activity and extracellular enzymes production. In contrast, negative priming, from decreased SOM mineralisation, would result in decreased microbial activity due to divergence to new more readily available nutrient sources, inhibition due to environmental changes such as toxicity and reduced mineral bioavailability due to their adsorption to newly introduced carbon (Zimmerman et al., 2011; Jindo et al., 2012).

Thus Zimmerman et al., (2011) reported a positive priming effect of grass biochar produced at 250°C and 400°C during a 90-day application to low organic carbon content soil, in particular. However, extending the incubation period to 250 and 500 days resulted in negative primed carbon mineralisation especially for supplementation with the hardwood-derived 525°C and 650°C biochars. Also, a positive priming effect was reported in an 87-day study probably due to the water soluble fraction of the low temperature (350°C) biochar (Luo et al., 2011). Although neither study identified the biotic or abiotic CO_2 source(s) used to measure the priming effect, Zimmerman et al., (2011) proposed that increased mineralisation of labile biochar and positive priming effects during early incubation would probably be a response of r-strategists.

Parallel to the reported biochar-driven positive priming effects, some studies have revealed neutral and negative priming. For example, pyrogenic carbon did not contribute to dissolved soil organic carbon concentration (Kuzyakov et al., 2009); pyrogenic organic carbon did not affect indigenous carbon or nitrogen mineralization (Santos et al., 2012); water soluble organic carbon and nitrogen were not affected (Jones et al., 2012); SOM degradation decreased (Jones et al., 2011); and water soluble organic carbon was reduced significantly (Gomez-Eyles et al., 2011).

Water soluble carbon sources, especially carbohydrates and phenols, can be utilized preferentially by soil microorganisms hence ecosystem functions could be affected negatively by their decreased bioavailability as a result of biochar addition (Jindo et al., 2012). Overall, potentially simultaneous biotic and abiotic induced mineralization (Lehmann and Sohi, 2008 as cited in Luo et al., 2011), plus complex SOM-soil microorganism-biochar interactions, make explaining the priming effect mechanisms difficult and necessitate robust experimental investigation.

Analysis of the rate of biochar mineralisation relative to bacterial biomass in compost did not show any increase in response to biochar addition during the first and sixth week (Jindo et al., 2012). Similar observations were made by Khodadad et al., (2011) where grass- and oak-derived biochars induced no taxa-specific changes, with the researchers proposing microbial utilization and mineralisation of refractory carbon as a potential mechanism. The non-stable relatively recalcitrant carbon fraction of biochar would not be readily usable by soil microorganisms (Masek et al., 2011). Therefore, utilization of refractory biochar-C concomitant with significant microbial response immediately following application, could signify the role of other factors for microbial utilization of biochar.

According to Khodadad et al., (2011), biochar addition to periodically burned soil showed higher bacterial counts and increased growth compared to unburned soil. This was attributed to potential of bacterial adaptation to pyrogenic carbon in the burned soil with the subsequent primed/increased mineralization of the newly added pyrogenic carbon. In contrast, increased mineralization of biochar carbon was reported in bare fallow soil. Thus Cross and Sohi (2011), produced biochars from sugarcane bagasse at different pyrolysis temperatures (350, 400 and 550°C) and different residence process times. These permutations resulted in different biochar-based labile carbon concentrations. Also, the different carbon mineralization rates (assessed by CO_2 flux) were attributed to differences soil microbial community degradative capacities for the labile carbon and soil types (bare fallow, arable and grassland). An increased microbial mineralisation of biochar carbon was observed in bare fallow soils.

The bioavailability of a new carbon source from the labile biochar-C in the nutrient poor fallow soil probably accounted for increased carbon mineralisation (Cross and Sohi, 2011).

The impacts of volatile organic compounds (VOCs) on soil-plant-microbial systems are economically, agriculturally and environmentally important. They, therefore, justify the need to elucidate the effects of biochar-derived VOCs on soil microbial community activities and function, as well as the mechanisms for VOC sorption by biochar. The concentrations of VOC from biochar differ depending on feedstock type and composition and biochar production methods and process conditions. Thus Li et al., (2011) observed 3,290 μgkg-1 naphthalene, which were produced during pyrolysis of apple wood sawdust. In a separate study, Spokas et al., (2011) reported ethylene as one of several VOCs and suggested its potential effects on the observed microbial responses. Also, since biochar can sorb volatile organic compounds from the surrounding environment, the researchers reported sorption of ≤20% naphthalene. Biochars produced from orange peel, paper mill waste and maize stover sorbed naphthalene hence recommendations were made for their use to remediate naphthalene contaminated sites (Chen et al., 2011; Oleszczuk et al., 2012). Naturally, this mandates detailed investigations on the catabolic microbial community responses to these biochars, including with and without magnetism, particularly before *in situ* applications.

Also, parallel analyses of bacterial responses to biochar and naphthalene have been made under different experimental conditions often with different outcomes. Therefore, comprehensive studies must be made to address the current critical knowledge gap on the microbial responses to biochar in the presence of naphthalene as a model contaminant. The findings can then ensure informed decision-making for biochar-based bioremediation of naphthalene/ PAH contaminated ecosystems.

CONCLUSION

Parallel critical literature analyses of bacterial response to biochar, effects of naphthalene on bacterial species/communities, and the interactions between biochar and naphthalene have highlighted the following: biochar affects bacterial strains and populations; naphthalene degrading bacteria respond to molecule concentration and bioavailability; and biochar can sorb naphthalene. Therefore, it can be expected that naphthalene catabolising microbial communities would be affected by biochar-based bioavailability. Several other factors would also determine the bacterial community response. For example, biochar can: be a nutrient source; induce soil organic matter mineralisation; alter soil characteristics such as pH, aeration, surface area, CEC, aggregation, density and moisture content; and absorb/adsorb enzymes. These would be dependent, subsequently, on the ecosystem, biochar type and process conditions (Lehmann and Joseph, 2009).

REFERENCES

Abouseoud, M., Yataghene, A., Amrane, A., Maachi, R. 2010. Effect of pH and salinity on the emulsifying capacity and naphthalene solubility of a biosurfactant produced by *Pseudomonas fluorescens*. *J. Hazard. Mater.* 180, 131-136.

Ahmad, M., Rajapaksha, A. U., Lim, J. E., Zhang, M., Bolan, N., Mohan, D., Vithanage, M., Lee, S. S., Ok, Y. S. 2014. Biochar as a sorbent for contamination management in soil and water. *Chemosphere* 99, 19-33.

Alberti, G., Amendola, V., Pesavento, M., Biesuz, R. 2012. Beyond the synthesis of novel solid phases: Review on modelling of sorption phenomena. *Coord. Chem. Rev.* 256, 28-45.

Alves, E. S., de Souza, S. R., Pedroso, A. N. V., Domingos, M. 2008. Potential of the Trad-MCN assay applied with inflorescences of *Tradescantia pallida* 'Purpurea' for evaluating air contamination by naphthalene. *Ecotoxicol. Environ. Saf.* 71, 717-721.

Anderson, C. R., Condron, L. M., Clough, T. J., Fiers, M., Stewart, A., Hill, R. A., Sherlock, R. R. 2011. Biochar induced soil microbial community change: Implications for biogeochemical cycling of carbon, nitrogen and phosphorus. *Pedobiologia* 54, 309-320.

Ania, C. O., Cabal, B., Pevida, C., Arenillas, A., Parra, J. B. Rubiera, F., Pis, J. J. 2007. Removal of naphthalene from aqueous solution on chemically modified activated carbons. *Water Res.* 41, 333-340.

Arulazhagan, P., Vasudevan, N. 2011. Biodegradation of polycyclic aromatic hydrocarbons by a halotolerant bacterial strain *Ochrobactrum* sp. VA1. *Mar. Pollut. Bull.* 62, 388-394.

Balachandran, C., Duraipandiyan, V., Balakrishna, K., Ignacimuthu, S. 2012. Petroleum and polycyclic aromatic hydrocarbons (PAHs) degradation and naphthalene metabolism in *Streptomyces* sp. (ERI-CPDA-1) isolated from oil contaminated soil. *Bioresour. Technol.* 112, 83-90.

Barret, M., Carrere, H., Delagadillo, L., Patureau, D. 2010. PAH fate during the anaerobic digestion of contaminated sludge: Do bioavailability and/or cometabolism limit their biodegradation? *Water Res.* 44, 3797-3806.

Bayard, R., Barna, L., Mahjoub, B., Gourdon, R. 2000. Influence of the presence of PAHs and coal tar on naphthalene sorption in soils. *J. Contam. Hydrol.* 46, 61-80.

Beesley, L., Moreno-Jiménez, E., Gomez-Eyles, J. L. 2010. Effects of biochar and greenwaste compost amendments on mobility, bioavailability and toxicity of inorganic and organic contaminants in a multi-element polluted soil. *Environ. Pollut.* 15, 2282-2287.

Beesley, L., Moreno-Jiménez, E., Gomez-Eyles, J. L., Harris, E., Robinson, B., Sizmur, T. 2011. A review of biochars' potential role in the remediation, revegetation and restoration of contaminated soils. *Environ. Pollut.* 159, 3269-3282.

Bruun, E. W., Hauggaard-Nielsen, H., Ibrahim, N., Egsgaard, H., Ambus, P., Jensen, P. A., Dam-Johansen, K. 2011. Influence of fast pyrolysis temperature on biochar labile fraction and short-term carbon loss in a loamy soil. *Biomass Bioenergy* 35, 1182-1189.

Burgos, W. D., Berry, D. F., Bhandari, A., Novak, J. T. 1999. Impact of soil-chemical interactions on the bioavailability of naphthalene and 1-naphthalenethol. *Water Res.* 33, 3789-3795.

Cai, J., Yang, S., Li, T. 2011. Logistic distributed activation energy model – Part 2: Application to cellulose pyrolysis. *Bioresour. Technol.* 102, 3642-3644.

Cao, X., Harris, W. 2010. Properties of dairy-manure-derived biochar pertinent to its potential use in remediation. *Bioresour. Technol.* 101, 5222-5228.

Cébron, A., Faure, P., Lorgeoux, C., Ouvard, S., Leyval, C. 2013. Experimental increase in availability of a PAH complex organic contamination from an aged contaminated soil: Consequences on biodegradation. *Environ. Pollut.* 177, 98-105.

Chen, B., Zhou, D., Zhu, L. 2008. Transitional adsorption and partition on nonpolar and polar aromatic contaminants by biochars of pine needles with different pyrolytic temperatures. *Environ. Sci. Technol.* 42, 5137-5143.

Chen, B., Chen, Z. 2009. Sorption of naphthalene and 1-naphthalenethol by biochars of orange peels with different pyrolytic temperatures. *Chemosphere* 76, 127-133.

Chen, B., Chen, Z., Lv, S. 2011. A novel magnetic biochar efficiently sorbs organic pollutants and phosphate. *Bioresour. Technol.* 102, 716-723.

Cross, A. and Sohi, S. P. 2011. The priming potential of biochar products in relation to labile carbon contents and soil organic matter status. *Soil Biol. Biochem.* 43, 2127-2134.

Di Gennaro, P., Moreno, B., Annoni, E., García-Rodríguez, S., Bestetti, G., Benitez, E. 2009. Dynamic changes in bacterial community structure and in naphthalene dioxygenase expression in vermicompost-amended PAH-contaminated soils. *J. Hazard. Mater.* 172, 1464-1469.

Dsikowitzky, L., Nordhaus, I., Jennerjahn, T. C., Khrycheva, P., Sivatharshan, Y., Yuwono, E., Schwarzbauer, J. 2011. Anthropogenic organic contaminants in water, sediments and benthic organisms of the mangrove-fringed Segara Anakan Lagoon, Java, Indonesia. *Mar. Pollut. Bull.* 62, 851-862.

Duku, M. H., Gu, S., Hagan, E. B. 2011. Biochar production potential in Ghana—A review. *Renewable and Sustainable Energy Reviews* 15, 3539-3551.

Feng, Y., Xu, Y., Yu, Y., Xie, Z., Lin, X. 2012. Mechanisms of biochar decreasing methane emission from Chinese paddy soils. *Soil Biol. Biochem.* 46, 80-88.

Filonov, A. E., Puntus, I. F., Karpov, A. V., Gaiazov, R. R., Kosheleva, I. A., Boronin, A. M. 1999. Growth and survival of *Pseudomonas putida* strains degrading naphthalene in soil model systems with different moisture levels. *Process Biochem.* 34, 303-308.

Glaser, B., Birk, J. J. 2012. State of the scientific knowledge on properties and genesis of Anthropogenic Dark Earths in Central Amazonia (*terra preta de Índio*). *Geochim. Cosmochim. Acta.*, 82, 39-51.

Gomes, N. C. M., Kosheleva, I. A., Abraham, W., Smalla, K. 2005. Effects of the inoculant strain *Pseudomonas putida* KT2442 (pNF142) and of naphthalene contamination on the soil bacterial community. *FEMS Microbiol. Ecol.* 54, 21-33.

Gomez-Eyles, J. L., Sizmur, T., Collins, C. D., Hodson, M. E. 2011. Effects of biochar and the earthworm *Eisenia fetida* on the bioavailability of polycyclic aromatic hydrocarbons and potentially toxic elements. *Environ. Pollut.* 159, 616-622.

Haritash, A. K., Kaushik, C. P. 2009. Biodegradation aspects of polycyclic aromatic hydrocarbons (PAHs): A review. *J. Hazard. Mater.* 169, 1-15.

Hennessee, C. T., Seo, J.-S., Alvarez, A. M., Li, Q. X. 2009. Polycyclic aromatic hydrocarbon-degrading species isolated from Hawaiian soils: *Mycobacterium crocinum* sp. nov., *Mycobacterium pallens* sp. nov., *Mycobacterium rutilum* sp. nov., *Mycobacterium rufum* sp. nov. and *Mycobacterium aromaticivorans* sp. nov. *Int. J. Syst. Evol. Microbiol.* 59, 378-387.

Hilscher, A., Knicker, H. 2011. Carbon and nitrogen degradation on molecular scale of grass-derived pyrogenic organic material during 28 months of incubation in soil. *Soil Biol. Biochem.* 43, 261-270.

Hwang, G., Park, S. R., Lee, C.-H., Ahn, I.-S., Yoon, Y.-J., Mhin, B. J. 2009 Influence of naphthalene biodegradation on the adhesion of *Pseudomonas putida* NCIB 9816-4 to a naphthalene-contaminated soil. *J. Hazard. Mater.* 172, 491-493.

Inyang, M., Gao, B., Yao, Y., Xue, Y., Zimmerman, A. R., Pullammanappallil, P., Cao, X. 2012. Removal of heavy metals from aqueous solution by biochars derived from anaerobically digested biomass. *Bioresour. Technol.* 110, 50-56.

Ji, X., Ripp, S. A., Layton, A. C., Sayler, G. S., Debruyn, J. M. 2013. Assessing long term effects of bioremediation: Soil Bacterial communities 14 years after polycyclic aromatic hydrocarbon contamination and introduction of a genetically engineered microorganism. *Bioremediat. Biodegrad.* 4, 8.

Jindo, K., Sánchez-Monedero, M. A., Hernández, T., García, C., Furukawa, T., Matsumoto, K., Sonoki, T., Bastida, F. 2012a. Biochar influences the microbial community structure during manure composting with agricultural wastes. *Sci. Total Environ.* 416, 476-481.

Jindo, K., Suto, K., Matsumoto, K., García, C., Sonoki, T., Sanchez-Monedero, M. A. 2012b. Chemical and biochemical characterisation of biochar-blended composts prepared from poultry manure. *Bioresour. Technol.* 36, 396-404.

Johnsen, A. R., Wick, L. Y., Harms, H. 2005. Principles of microbial PAH-degradation in soil. *Environ. Pollut.* 133, 71-84.

Jones, D. L., Rousk, J., Edwards-Jones, G., DeLuca, T. H., Murphy, D. V. 2012. Biochar-mediated changes in soil quality and plant growth in a three year field trial. *Soil Biol. Biochem.* 45, 113-124.

Kafilzadeh, F., Rafiee, S., Tahery, Y. 2011. Evaluation of Bioremediation of naphthalene using native bacteria isolated from oil contaminated soils in Iran. *Annals of Biological Research* 2, 610-616.

Khodadad, C. L. M., Zimmerman, A. R., Green, S. J., Uthandi, S., Foster, J. S.,2011. Taxa-specific changes in soil microbial community composition induced by pyrogenic carbon amendments. *Soil Biol. Biochem.* 43, 385-392.

Krell, T., Lacal, J., Guazzaroni, M. E., Busch, A., Silva-Jiménez, H., Fillet, S., Reyes-Darías, J. A., Muñoz-Martínez, F., Rico-Jiménez, M., García-Fontana, C., Duque, E., Segura, A., Ramos, J.-L. 2012. Responses of *Pseudomonas putida* to toxic aromatic carbon sources. *J. Biotechnol.* 160, 25-32.

Lee, K., Park, J.-W., Ahn, I.-S. 2003. Effect of additional carbon source on naphthalene biodegradation by *Pseudomonas putida* G7. *J. Hazard. Mater.* 105, 157-167.

Lehmann, J., Rillig, M. C., Thies, J., Masiello, C. A., Crowley, D. 2011. Biochar effects on soil biota – A review. *Soil Biol. Biochem.* 43, 1812-1836.

Lehmann, J., Joseph, S. 2009. *Biochar for Environmental Management: Science and Technology*, Earth Scan Publishers.

Li, D., Hockaday, W. C., Masiello, C. A., Alvarez, P. J. J., 2011. Earthworm avoidance of biochar can be mitigated by wetting. *Soil Biol. Biochem.* 43, 1732-1737.

Li, J. L., Chen, B. H. 2009. Surfactant-mediated biodegradation of polycyclic aromatic hydrocarbons. *Materials* 2, 76-94.

Lin, C., Gan, L., Chen, Z.-L. 2010. Biodegradation of naphthalene by strain *Bacillus fusiformis* (BFN). *J. Hazard. Mater.* 182, 771-777.

Lou, L., Wu, B., Wang, L., Luo, L., Xu, X., Hou, J., Xun, B., Hu, B., Chen, Y. 2011. Sorption and ecotoxicity of pentachlorophenol polluted sediment amended with rice-straw derived biochar. *Bioresour. Technol.* 102, 4036-4041.

Lu, X., Zhang, T., Han-Ping Fang, H., Leung, K. M. Y., Zhang, G. 2011. Biodegradation of naphthalene by enriched marine denitrifying bacteria. *Int. Biodeterior Biodegradation* 65, 204-211.

Lucas, N., Bienaime, C., Belloy, C., Queneudec, M., Silvestre, F., Nava-Saucedo, J.-E. 2008. Polymer biodegradation: Mechanisms and estimation techniques – A review. *Chemosphere* 73, 442-429.

Luo, Y., Durenkamp, M., De Nobili, M., Brookes, P. C. 2011. Short term soil priming effects and the mineralisation of biochar following its incorporation to soils of different pH. *Soil Biol. Biochem.* 43, 2304-2314.

Ma, L., Zhang, J., Han, L., Li, W., Xu, L., Hu, F., Li, H., 2012. The effects of aging time on the fraction distribution and bioavailability of PAH. *Chemosphere* 86, 1072-1078.

Mao, J., Luo, Y., Teng, Y., Li, Z. 2012. Bioremediation of polycyclic aromatic hydrocarbon-contaminated soil by a bacterial consortium and associated microbial community changes. *Int. Biodeterior. Biodegradation* 70, 141-147.

Martin, F., Torelli, S., Le Paslier, D., Barbance, A., Martin-Laurent, F., Bru, D., Geremia, R., Blake, G., Jouanneau, Y. 2012. Betaproteobacteria domaninace and diversity shifts in the bacterial community of a PAH-contaminated soil exposed to phenanthrene. *Environ. Pollut.* 162, 345-353.

Mašek, O., Brownsort, P., Cross, A., Sohi, S. 2011. Influence of production conditions on the yield and environmental stability of biochar. *Fuel* 103, 151-155.

Matovic, D. 2011. Biochar as a viable carbon sequestration option: Global and Canadian perspective. *Energy* 36, 2011-2016.

Megharaj, M., Ramakrishnan, B., Venkateswarlu, K., Senthunathan, N., Naidu, R. 2011. Bioremediation approaches for organic pollutants: A critical perspective. *Environ. Int.* 37, 1362-1375.

Menzie, C. A., Potochi, B. B., Santodonato, J. 1992. Exposure to carcinogenic PAHs in the environment. *Environ. Sci. Technol.* 26, 1278-1284.

Mittal, M., Rockne, K. J. 2009. Naphthalene and phenanthrene sorption to very low organic content diatomaceous earth: Modelling implications for microbial bioavailability. *Chemosphere* 74, 1134-1144.

Mrozik, A., Łabużek, S., Piotrowska-Seget, Z. 2005. Changes in fatty acid composition in *Pseudomonas putida* and *Pseudomonas stutzeri* during naphthalene degradation. *Microbiol. Res.* 160, 149-157.

Mrozik, A., Piotrowska-Seget, Z., Łabużek, S. 2004. Changes in whole cell-derived fatty acids induced by naphthalene in bacteria from genus *Pseudomonas*. *Microbiol. Res.* 159, 87-95.

Mrozik, A., Piotrowska-Seget, Z., Łabużek, S. 2003. Bacterial degradation and bioremediation of polyaromatic hydrocarbons. *Polish Journal of Environmental Studies* 12, 15-25.

Muckian, L., Grant, R., Doyle, E., Clipson, N. 2007. Bacterial community structure in soils contaminated by polycyclic aromatic hydrocarbons. *Chemosphere* 68, 1535-1541.

Nag, S. K., Kookana, R., Smith, L., Krull, E., Macdonald, L. M., Gill, G. 2011. Poor efficacy of herbicides in biochar-amended soils as affected by their chemistry and mode of action. *Chemosphere* 84, 1572-1577.

Nguyen, B. T., Lehmann, J. 2009. Black carbon decomposition under varying water regimes. *Org. Geochem.* 40, 846-853.

Nichols, N. N., Harwood, C. S. 2000. An aerotaxis transducer gene from *Pseudomonas putida*. *FEMS Microbiol. Lett.* 182, 177-183.

Ohsowski, B. M., Klironomos, J. N., Dunfield, K. E., Hart, M. M. 2012. The potential of soil amendments for restoring severely disturbed grasslands. *Appl. Soil Ecol.* 60, 77-83.

Oleszczuk, P., Hale, S., Lehmann, J., Cornelissen, G. 2012. Activated carbon and biochar amendments decrease porewater concentrations of polycyclic aromatic hydrocarbons (PAHs) in sewage. *Bioresour. Technol.* 111, 84-91.

O' Neill, B., Grossman, J., Tsai, M. T., Gomes, J. E., Lehmann, J., Peterson, J., Neves, E., Thies, J. E. 2009. Bacterial community composition in Brazilian Anthrosols and adjacent soils characterized using culturing and molecular identification. *Microb. Ecol.* 58, 23-35.

Ono, A., Miyazaki, R., Sota, M., Ohtsubo, Y., Nagata, Y., Tsuda, M. 2007. Isolation and characterization of naphthalene-catabolic genes and plasmids from oil-contaminated soil by using two cultivation-independent approaches. *Appl. Microbiol. Biotechnol.* 74, 501-510.

Panwar, N. L., Kothari, R., Tyagi, V. V. 2012. Thermo chemical conversion of biomass – Eco friendly energy routes. *Renew. Sust. Energ. Rev.* 16, 1801-1816.

Patel, V., Jain, S., Madamwar, D. 2012. Naphthalene degradation by bacterial consortium (DV-AL) developed from Alang-Sosiya ship breaking yard, Gujarat, India. *Bioresour. Technol.* 107, 122-130.

Pathak, H., Kantharia, D., Malpani, A., Madamwar, D. 2009. Naphthalene degradation by *Pseudomonas* sp. HOB1: *In vitro* studies and assessment of naphthalene degradation efficiency in simulated microcosms. *J. Hazard. Mater.* 166, 1466-1473.

Pawar, A. N., Ugale, S. S., More, M. G., Kokani, N. F., Khandelwal, S. R. 2013. Biological degradation of naphthalene: A new Era. *Bioremediat. Biodegrad.* 4, 7.

Peng, X., Ye, L. L., Wang, C. H., Zhou, H., Sun, B. 2011. Temperature- and duration-dependent rice straw-derived biochar: Characteristics and its effects on soil properties of an Ultisol in southern China. *Soil Tillage Research* 112, 159-166.

Pratt, K., Moran, D. 2010. Evaluating the cost-effectiveness of global biochar mitigation potential. *Biomass Bioenergy* 34, 1149-1158.

Puglisi, E., Cappa, F., Fragoulis, G., Trevisan, M., Del Re, A. A. M. 2007. Bioavailability and degradation of phenanthrene in compost amended soils. *Chemosphere* 67, 548-556.

Puntus, I. P., Sakharovsky, V. G., Filonov, A. E., Boronin, A. M. 2005. Surface activity and metabolism of hydrocarbon-degrading microorganisms growing on hexadecane and naphthalene. *Process Biochem.* 40, 2643-2648.

Purwaningsih, I. S., Hill, G. A., Headley, J. V. 2004 Mass transfer and bioremediation of naphthalene particles in a roller bioreactor. *Water Res.* 38, 2034-3027.

Qui, Y., Zheng, Z., Zhou, Z., Sheng, G. D. 2009. Effectiveness and mechanisms of dye adsorption on a straw-based biochar. *Bioresour. Technol.* 100, 5348-5351.

Reddy, K. R., Xie, T., Dastgheibi, S. 2014. Evaluation of biochar as a potential filter media for the removal of mixed contaminants from urban storm runoff. *J. Environ. Eng.* 140.

Ron, E. Z., Rosenberg, E. 2002. Biosurfactants and oil bioremediation. *Curr. Opin. Biotechnol.* 13, 249-252.

Ruthiraan, M., Mubarak, N. M., Thines, R. K., Abdullah, E. C., Sahu, J. N., Jayakumar, N. S., Ganesan, P. 2015. Comparative kinetic study of functionalized carbon nanotubes and magnetic biochar for removal of Cd^{2+} ions from wastewater. *Korean J. Chem. Eng.* 32, 446-457.

Santos, F., Torn, M. S., Bird, J. A. 2012. Biological degradation of pyrogenic organic matter in temperate forest soils. *Soil Biol. Biochem.* 51, 115-124.

Schreiberová, O., Hedbavná, P., Čejková, A., Masák, J. 2012. Effect of surfactants on the biofilm of *Rhodococcus erythropolis*, a potent degrader of aromatic pollutants. *N. Biotechnol.* 15, 62-68.

Seo, J., Keum, Y., Li, Q. X. 2009. Bacterial degradation of aromatic compounds. *Int. J. Environ. Res. Public Health.* 6, 278-309.

Shi, Z., Tian, L., Zhang, Y. 2010. Molecular biology approaches for understanding microbial polycyclic aromatic hydrocarbons (PAHs) degradation. *Acta Ecologica Sinica*, 30, 292-295.

Shimada, K., Itoh, Y., Washio, K., Morikawa, M. 2012. Efficacy of forming biofilms by naphthalene degrading *Pseudomonas stutzeri* T102 toward bioremediation technology and its molecular mechanisms. *Chemosphere* 87, 226-233.

Singer, A. C., Crowley, D. E. Thompson, I. P. 2003. Secondary plant metabolites in phytoremediation and biotransformation. *Trends Biotechnol.* 21, 123-130.

Singleton, D. R., Powell, S. N., Sangaiah, R., Gold, A., Ball, L. M., Aitken, M. D. 2005. Stable-isotope probing of bacteria capable of degrading salicylate, naphthalene or phenanthrene in a bioreactor treating contaminated soil. *Appl. Environ. Microbiol.* 71, 1202-1209.

Smith, J. L., Collins, H. P., Bailey, V. L. 2010. The effect of young biochar on soil respiration. *Soil biol. Biochem.* 42, 2345-2347.

Sobisch, T., Heß, H., Niebelschütz, H., Schmidt, U. 2000. Effect of additives on biodegradation of PAH in soils. *Colloids Surf. A Physicochem. Eng. Asp.* 162, 1-14.

Sohi, S. P., Krull, E., Lopez-Capel, E., Bol, R. 2010. A Review of biochar and its use and function in soil. *Adv. Agron.* 105, 47-82.

Song, Y., Wang, F., Bian, Y., Kengara, F. O., Jia, M., Xie, Z., Jiang, X. 2012. Bioavailability assessment of hexachlorobenzene in soil as affected by wheat straw biochar. *J. Hazard. Mater.* 217-218, 391-397.

Spokas, K. A., Novak, J. M., Stewart, C. E., Cantrell, K. B., Uchimiya, M., DuSaire, M. G., Ro, K. S. 2011. Qualitative analysis of volatile organic compounds on biochar. *Chemosphere* 85, 869-882.

Steinbeiss, S., Gleixner, G., Antonietti, M. 2009. Effect of biochar amendment on soil carbon balance and soil microbial activity. *Soil Biol. Biochem.* 41, 1301-1310.

Torsvik, V., Øvreås, L. 2002. Microbial diversity and function in soil: From genes to ecosystems. *Curr. Opin. Microbiol.* 5, 240-245.

Uhlik, O., Wald, J., Strejcek, M., Musilova, L., Ridl, J., Hroudova, M., Vlcek, C., Cardenas, E., Mackova, M., Macek, T. 2012. Identification of bacteria utilizing biphenyl, benzoate and naphthalene in long-term contaminated soil. *Plos One* 7(7), e40653.

Valverde, A., González-Tirante, M., Medina-Sierra, M., Santa-Regina, I., García-Sánchez, A., Igual, J. M. 2011. Diversity and community structure of culturable arsenic-resistant bacteria across a soil arsenic gradient at an abandoned tungsten–tin mining area. *Chemosphere* 85, 129-134.

Verheijen, F. G. A., Jeffery, S., Bastos, A. C., van der Velde, M., Diafas, I. 2009. *Biochar Application to Soils - A Critical Scientific Review of Effects on Soil Properties, Processes and Functions.* EUR 24099 EN, Office for the Official Publications of the European Communities, Luxembourg, p. 149.

Wang, S., Gao, B., Zimmerman, A. R., Li, Y., Ma, L., Haris, W. G., Migiliaccio, K. W. 2015. Removal of arsenic by magnetic biochar prepared from pinewood and natural hematite. *Bioresour. Technol.* 175, 391-395.

Warnock, D. D., Mummey, D. L., McBride, B. Major, J., Lehmann, J., Rillig, M. C. 2010. Influences of non-herbaceous biochar on arbuscular mycorrhizal fungal abundances in roots and soils: Results from growth-chamber and field experiments. *Appl. Soil Ecol.* 46, 450-456.

Xu, T., Lou, L., Luo, L., Cao, R., Duan, D., Chen, Y. 2012. Effect of bamboo biochar on pentachlorophenol leachability and bioavailability in agricultural soil. *Sci. Total Environ.* 414, 727-731.

Xu-Xiang, Z., Shu-Pei, C., Cheng-Jun, Z., Shi-Lei, S. 2006. Microbial PAH-degradation in soil: Degradation pathways and contributing factors. *Pedosphere* 16, 555-565.

Yao, Y., Gao, B., Inyang, M. 2011. Biochar derived from anaerobically digested sugar beet tailings: Characterization and phosphate removal potential. *Bioresour. Technol.* 102, 6273-6278.

Yan, L., Kong, L., Qu, Z., Li, L., Shen, G. 2015. Magnetic biochar decorated with ZnS nanocrytals for Pb (II) removal. *ACS Sustain. Chem. Eng.* 3, 125-132.

Yen, K. M., Gunsalus, I. C. 1985. Regulation of naphthalene catabolic genes of plasmid NAH7. *J. Bacteriol.* 162, 1008-1013.

Zavalloni, C., Alberti, G., Biasiol, S., Vedove, G. D., Fornasier, F., Liu, J., Peressotti, A. 2011. Microbial mineralization of biochar and wheat straw mixture in soil: A short-term study. *Appl. Soil Ecol.* 50, 45-51.

Zeinali, M., Vossoughi, M., Ardestani, S. K. 2008. Naphthalene metabolism in *Nocardia otitidiscaviarum* strain TSH1, a moderately thermophilic microorganism. *Chemosphere* 72, 905-909.

Zhang, J., Zeng, J., He, M. 2009. Effects of temperature and surfactants on naphthalene and phenanthrene sorption by soil. *J. Environ. Sci.* 21, 667-674.

Zhang, M., Liang, S., Shen, X., Guo, X., Tao, S., Xing, B., Wang, X. 2014. Characterization of nitrogen-rich biomaterial-derived biochars and their sorption for aromatic compounds. *Environ. Pollut.* 195, 84-90.

Zheng, W., Guo, M., Chow, T., Bennett, D. N., Rajagopalan, N. 2010. Sorption properties of greenwaste biochar for two triazine pesticides. *J. Hazard. Mat.* 181, 121-126.

Zimmerman, A. R., Gao, B., Ahn, M. Y. 2011. Positive and negative carbon mineralization priming effects among a variety of biochar-amended soils. *Soil Biol. Biochem.* 43, 1169-1179.

In: Soil Management
Editor: Manuel Esteban Lucas Borja

ISBN: 978-1-63483-274-8

Chapter 7

BIOFERTILIZER APPLICATION IN SOIL

Gabriela Mercuri Quiterio[1], Cassiana Coneglian[2], Ederio Dino Bidoia[1], Paulo Renato Matos Lopes[1], Ana Paula Justiniano Régo[1], Renato Nallin Montagnolli[1], Jaqueline Matos Cruz[1], Elis Marina Turini Claro[1], José Rubens Moraes Júnior[1] and Artur Blikstad Mauro[1]

[1]UNESP – São Paulo State University, Bioscience Institute, Rio Claro, São Paulo, Brazil

[2]UNICAMP – University of Campinas, Department of Technologies for Environmental Sanitation, Limeira, São Paulo, Brazil

ABSTRACT

A strong demand for healthy food free of chemical residues from pesticides used in pest control, diseases and invasive plants exists in our modern society. In addition, there is a permanent concern about environmental contamination by such chemical residues. Hence proper handling of chemical substances have been prioritized. It is also preferable the usage of natural substances and/or molecules of higher specificity and lower environmental impact.

An alternative and promising mechanism for increasing crop productivity is the use of biotechnological tools, especially the biological processes that occur in soil system/plant carried out by microorganisms, establishing the basis on which the agro-ecological system is based. In this context, biofertilizers arise as a sustainable alternative due to the lowest environmental impact and the cost-effective production values.

The biofertilizer is a liquid organic fertilizer and is obtained from the organic compound dissolved in water, followed by fermentation or digestion by microorganisms. It is a rich organic compound mixture of nutrients and microbial biomass able to supply the soil biota with substrates, thus assisting in the metabolic activity of the local community.

The process of obtaining the biofertilizer results in two phases: a solid that can be used as organic fertilizer in soil, and the liquid phase used as foliar fertilizer and pest control. These organic fertilizers are bioactive compounds, terminal residues of fermentation of organic compounds, containing living or latent microorganism cells.

Biofertilizers are produced in biodigesters by aerobic and/or anaerobic fermentation of organic matter and, therefore, its constitution is rich in enzymes, antibiotics, vitamins, toxins, phenols, esters and organic acids. Thus, due to the high concentration and diversity of nutrients in its composition, biofertilizers become an important tool in accelerating xenobiotic degradation of microbial processes in soil.

Keywords: biofertilizer, microorganisms, biodegradation

Introduction

Brazil's agribusiness has been growing due to increased production capacity and investment in advanced technology. In recent years, few countries had growth in international agribusiness trade as Brazil, which is a world leader in production and export of various agricultural products like soybeans, coffee, fruit juices, sugar and alcohol.

This increase in industrial activity and population growth caused the environmental pollution problem to reach alarming levels. In addition to contamination by poorly biodegradable waste, it adds problem of industrial waste and especially agro-industrial residues, such as sugarcane vinasse or stillage from ethanol production (Medeiros et al., 2003). This is one of the major problems associated with the alcohol industry regarding the large volume of residues produced daily and their high pollution potential when discarded in large amounts in soil and water bodies. Due to this exceeding volume and the generation of sugarcane vinasse there is a growing concern about the disposal and treatment of this effluent. These problems can be solved by bioremediation process, which is based on the degradation of contaminants in soil by the activity of microorganisms present or added at the site of contamination (BERNOTH et al., 2000).

To assist in the bioremediation process, the biofertilizers are being studied as a sustainable alternative fertilization method due to their low environmental impact and high cost-effectiveness (DELEITO et al., 2005). In bioremediation, the biofertilizer can be used as an alternative to biodegradation of xenobiotics, because it provides nutrients to microorganisms responsible of treatment (Cutright and Lee, 1994).

The biofertilizer can be defined as a substance that contains live microorganisms which when applied to the seeds, plants, surfaces or soil can colonize rhizosphere of the plant or the inner and promotes growth by providing or increasing the availability of primary nutrients the host plant (Vessey, 2003). The biofertilizers in plants are effective in the control of pests and diseases, growth acceleration and nutritional status (Santos, 1992; Bettiol et al., 1998; Penteado, 2004). In soil, they can contribute to physical improvement and promote the production of humic substances, which have significant importance in soil fertility with a positive impact on agricultural production (Delgado et al., 2002).

Biofertilizers

Biofertilizer is a liquid organic fertilizer derived from aerobic or anaerobic composting. Microorganisms act in complexation and decomposition of a mixture of organic material (manure, for example), minerals, and water (D'Andrea, 2010). According to Vessey (2003), fertilizers and biofertilizers are distinct compounds. Fertilizers are chemical compounds that increase soil fertility, whereas biofertilizers are organic compounds in which living microorganisms contained there improve the nutritional status of plants, promoting vegetable growth.

The process to obtain biofertilizer results in two phases, a solid one that can be used as organic fertilizer in soil and a liquid phase that can be used as foliar fertilizer and pest control. It is produced in biodigesters by aerobic and/or anaerobic fermentation of organic matter and there is no standard formula for its production. The fermentation process is complex and microorganisms pass through four distinct phases of cell growth: lag, exponential growth, stationary phase and cell death (D'Andrea, 2010).

The method of obtaining the biofertilizer described by Santos (1992) consists of the fermentation during thirty days in a closed system, with no air, of a mixture of fresh water and cattle manure in the proportion of 50% v/v. To obtain an anaerobic system, the mixture is placed in a plastic container of 200L, leaving a gap of 15 to 20 cm at its inside; closes hermetically and adapts a hose, immersing the other end in a container with water to the gas outlet. The product should not be stored for a long time, in order not to change its characteristics. If not fully utilized, the biofertilizer can be stored for a period of thirty days.

Another biofertilizer used is produced through aerobic digestion. To this, 20 L of fish offal, 10 kg of rice bran, 10 kg of wheat bran, 7 kg of bone meal, 2 kg of corn flour, 2 kg of flour and 5 kg of molasses are placed in a tank with 400 liters of water and mixed for five to ten minutes, twice or three times daily, or oxygen is pumped with the aid of an aquarium pump for forty days (Furlanetto et al., 1996). It is noteworthy also that, in all ways to prepare the biofertilizer, it is possible to make use of a continuous process, adding different products to keep it with high microbial activity. Thus, the biofertilizer has a constitution rich in enzymes, vitamins, toxins, phenols, organic acids and, esters. These substances are sources of macronutrients such as nitrogen, phosphorus, potassium, calcium, magnesium and micronutrients such as copper and zinc (Machado and Machado, 2009). In addition to its mineral composition, the biological potency of biofertilizer is expressed by the large amount of microorganisms responsible for the release of metabolites, including antibiotics and hormones.

Currently the use of biofertilizers has contributed to supplementation of nutrients and pest control in different cultures, whether irrigation system is applied to the soil or sprayed on plants (Souza and Resende, 2003). Their usage started back in the early 80s by Emater-Rio, belonging to the Department of Agriculture and Livestock of the State Rio de Janeiro, Brazil, in coffee and cane sugar plantations, washed down into the pits to make complementation and nutritional aid for irrigation since it was highly diluted. In 1985, the observation units in rubber, coffee and passion fruit began. The results achieved by spraying liquid biofertilizer at 20% dilution with water showed the reduction of plant pathogens and pests attack and increased productivity due to the balance of the ecosystem of the sprayed crops (Santos, 1991).

The effects of biofertilizer to control pests and diseases have been well evidenced. Fungistatic and bacteriostatic effects, these caused mainly by the presence of *Bacillus subtilis*, originating in the rumen of cattle have been observed (BETTIOL and STADNIK, 2001). Another beneficial effect was observed for Deleito et al., (2005) using the Agrobio biofertilizer on the development of pepper seedlings, where there was a reduction in incidence of bacterial spot. In the same study, the biofertilizers presented in a concentration above 5%, bacteriostatic effect on *Xanthomonas vesicatoria* in tests in vitro. In addition to antibiotic action, Barbosa and Medeiros (2007) cite the systemic resistance induction of the plant as the main biofertilizer mechanisms against plague.

The effects of biofertilizer applied in soil are the increase in organic matter content, activating the intrinsic local microbiota, improving its structure, aeration, water infiltration. Also, enhanced physical and chemical properties such as reduction of acidity, due to the ability of biofertilizer to retain bases, allows the formation of organic complexes and the production of negative charges (Galbiatti et al., 1996).

Ceretta et al., (2003) and Queiroz et al., (2004) concluded that the use of liquid fertilizer of pigs manure resulted in addition of large quantities of nutrients in soil, reflected in increased levels of phosphorus, calcium and magnesium in natural grass areas. Besides, there was improvement of fertility for plants growth. Damatto Junior et al., (2006) evaluated the effects of different doses of organic compost based of sawdust and manure, found increased pH, organic matter, phosphorus, calcium, sum of bases, cation exchange capacity (CEC) and base saturation soil. Increased levels of nitrogen, phosphorus, potassium, calcium, and manganese in the soil were observed by Alves et al., (2009) and considerable concentration of micronutrients such as boron, copper, chlorine, iron, molybdenum, manganese and zinc were observed by Tanaka et al., (2003) due to the provision of biofertilizer.

The effect of biofertilizers in plant nutrition is also observed by Kozen and Alvarenga (2005) had an increase in production of fodder maize and maize grain, under isolated or combined application of biofertilizers with chemical fertilizers. Ribeiro et al., (2000) concluded that organic matter provided the organic manure efficiently in the production of chili, increasing their productivity. Vidigal et al., (2010) showed that the application of organic compound based on solid swine manure is sufficient for the production of onion bulbs with high quality and increased productivity. In a study, Sediyama et al., (2009) found that application of swine biofertilizer provides increase in leaf nitrogen content, calcium, and magnesium and reduction in phosphorus and sulfur content, without loss in production for okra plant.

Biofertilizer and Bioremediation

Anthropogenic activities that lead to contaminants provisions may be harmful to soil, which may affect the microbial populations and their activities in different functional levels. The degradation of a substance can be further promoted by physical process such as heat or ultraviolet irradiation, by chemical reactions resulted of pH or certain substances in the environment action, and biological processes through enzymes of animals, plants, and microorganisms (Atlas and Bartha, 1993).

Bioremediation can be defined as the process by which living organisms (bacteria, fungi, plants and worms) degrade contaminants or become then into innocuous compounds such as carbon dioxide and water or less toxic forms (Paul et al., 2005 ; Vidali, 2001; USEPA, 2005). In soils, the microbiota is primarily responsible for the decomposition of organic waste, the cycling of nutrients and the flow of energy, influencing both the transformation of organic matter, as in carbon storage and minerals (Millioli et al., 2008). Bacteria are efficient in the degradation of organic contaminants in soils and have a number of characteristics that contribute to their use, such as rapid growth and metabolism, genetic plasticity and relatively quick adjustment ability to changes in environmental conditions (Baker and Herson, 1994).

The different bioremediation strategies generally aim to increase the microbial population creating environmental conditions favorable to their development (Silva et al., 2004). There are various techniques for *in situ* and *ex situ* bioremediation.

In-Situ Bioremediation

Accelerated Natural Attenuation (Biostimulation)

The accelerated natural attenuation technique (ANA) is a bioremediation option in-situ in which search, in especially; instigate the growth of native microbial population. This occurs, fundamentally, by the condition of the natural habitat of these microorganisms. Therefore, ANA, the basic idea is to condition the contaminated local, to accelerate the natural attenuation process, stimulating, particularly, the development of autochthonous microorganisms (EPA, 2004a).

For this bioremediation technique be efficiently, the first requirement to be met is the need of microorganisms present has ability to biodegrade the target pollutant. ANA, when used in a controlled manner, is able to accelerate natural process and reduce the time of bioremediation in site, and for this, generally it is added one or more of the following compounds: oxygen, hydrogen and nutrients. The addition of nutrients and optimization of environmental conditions allow increasing the growth rate and the metabolic activities of microorganisms, resulting in increased speed and percentage of biodegradation. In this context, biostimulation can offer lower cost compared to other bioremediation techniques (EPA, 2004a).

Bioaugmentation

Characterized by increase of microbiota by inoculating non-native microorganisms, not belonging to the site. The use of bioaugmentation in soils, basically, is justified by the need for rapid biodegradation of the pollutant compound and reducing the adjustment period that usually precedes the degradation process by indigenous microorganisms. However, this technique depends on the authorization of government agencies and environmental enforcement agencies. The approval of bioaugmentation as bioremediation strategy depends on several factors, among which, the types of microorganisms used and the application area.

The failure of the introduction of microorganisms may be related to some factors, such as nutrient limitation, elimination by predation, the concentration of very low organic pollutant to support the development and even environmental factors unfavorable to the growth of these microorganisms (Alexander, 1999; SIMS et al., 1992). In addition, the introduction of non-native microorganisms can result in considerable risks to the environment. Thus, in the light of risks, for example, in the use of biotechnology products, it is extremely important that these products are properly assessed before its application to, then, promote the degradation of pollutants in waste treatment systems or contaminated sites. This procedure should be performed carefully, seeking its use as safe and effective practice of environmentally and health view.

Ex-Situ Bioremediation

Biopiles

Ex situ bioremediation can be made by Biopiles technique, which is a biological processes under aerobic conditions. This bioremediation technique basically involves the disposal of contaminated materials in called biopiles hills. For this technique, initially the soil is excavated and then prepared and placed in biopiles, where it is done stimulation of microbial activity through aeration, nutrients addiction and increased soil moisture, in order to promote biodegradation of the contaminants of interest.

The biopiles are built on an impermeable base to reduce leachate migration potential of surface layer to subsurface soil. Moreover, an impermeable membrane to prevent release of volatile and semi-volatile contaminants from the soil to the atmosphere, as well as to protect the soil from natural weathering such as wind and rain covers them. The materials to be treated must be subjected to a previous study, for attesting the treatability, by conducting physical-chemical and biological assays and further optimization of laboratory conditions. The purpose of these studies were to evaluate the type and concentration of pollutants, the presence of possible degradation inhibitors, the structure of the material, the moisture optimum levels of nutrients and oxygen, temperature and other factors that influence the process (EPA, 2004b).

The treatment by biopiles endue as main advantage the possibility of greater control of the physical, biological and chemical conditions of contaminated matrix and, pursuant to, obtaining faster reaction kinetics and the reduction of the total time for contaminant degradation compared to natural attenuation process.

However, a major limitation is related to the removal of contaminated soil, in order that treatment process is made ex-situ. Even though this scenario allows the removal of the main sources of contamination in unsaturated zone, this practice should be made with caution. The removal may provide an increased risk of human exposure, such as inhalation of vapours and contaminated particulate materials (during the excavation, storage, transport and redeployment of contaminated soils).

Regardless of the bioremediation technique used, numerous factors determine its success, like: the characteristic of residue (as concentration and persistence), the selection of correct bioremediation technology, the use of analytical methods suitable for determining the type and extent of contamination and the presence of optimal microbiological conditions, such as essential nutrients for microbial growth (HUESEMANN, 1994). Thus, growth of microbial population often becomes necessary to add nutrients in the soil. Nutrients used in bioremediation can be grouped into three categories: inorganic soluble in water, oleophilic formulations and those called slow release, with slow dissolution of components. Soluble products are represented by a wide variety of agricultural fertilizers with different ratios of compounds containing nitrogen, phosphorus and potassium. The advantages of using these products are market availability and low cost (PRINCE, 1993; ATLAS, 1998).

The biofertilizers are, due to their nutritional value, an alternative for use in biological sewage treatment plants, since, as explained by Hammer and Hammer Júnior (2004), the biologically treated effluent should contain sufficient concentrations of carbon, nitrogen, phosphorus and trace nutrients to ensure adequate growth of microbial culture.

Effects of Sugarcane Vinasse on Environment

The vinasse, also known as stillage, is the residue after fractional distillation of the juice of cane sugar fermented to obtain ethanol (ethyl alcohol). For each litter of ethanol produced, from 8 to 18 litters of vinasse are obtained as residue.

The vinasse is composed mainly of water and solid salts in suspension or soluble, and the water content is between 94 and 96%. Vinasse is an effluent with a high fertilizer value but a high pollution power about a hundred times greater than sewage, which is due its richness in organic and mineral matter, with high Chemical Oxygen Demand (COD) and Biochemical Oxygen Demand (BOD), hence its increased pollutant potential. Besides pollution by organic substances, other physical aspects of the vinasse such as is its low pH (4.0 to 4.8) and high temperature result in a chemical pollutant property (expressed by high corrosive effect) and physical pollution (as heat). It can replace the nutrients found in mineral fertilization, in potassium followed by calcium, magnesium, phosphorus, manganese and organic nitrogen (Junqueira et al., 2009).

Because of its richness in organic matter, the application of vinasse in soil fertilization has given good results over years (Freire and Cortez, 2000). Biofertilizers can promote soil fertility, as it increases availability of some elements to plants, raising pH; increasing availability of certain ions; increased cation exchange capacity; increased water-holding capacity and improve the physical structure of soil. The stillage should be seen also as an agent of microbiota population boost and microbial activity growth in soil (Silva and Ribeiro, 1998).

However, the amount of vinasse fertigated should not exceed ions retention capacity of soil; it means that dosages depend on the characteristics of soil. When used in unbalanced proportions cause harmful effects to soils and plants.

As exacerbated effects of vinasse application in soil can be expected the increase in salt concentration and the potential risk of salinity over the years. The salinization consequences are: soil disruption, increase in bulk density and water retention and reduced water infiltration by excess sodium ions. These changes cause difficulties for plants to absorb water, specific ions toxicity and salts interference in physiological processes, thereby reducing plants growth and development (Rhoades et al., 2000).

Case Study

In a study conducted at Estate University of Campinas - UNICAMP, Campinas, São Paulo, Brazil, was verified the effect of addition of Microgeo, patented product (# PI0207342 A2 0) according to D'Andrea (2002), biofertilizer commercial use on soil microbial activity, as assessed by quantifying the amount of fungi and heterotrophic bacteria and bioremediation of sugarcane vinasse when added to soil.

Table 1. Quantification of bacteria and fungi of pure soil and soil with the addition of vinasse and/or biofertilizer after biodegradation process in respirometric test during the 90 days, evaluated in colony forming units per gram of soil (CFU g^{-1})

SAMPLE	BACTERIA	FUNGUS
SC	6.4×10^5	1.0×10^6
VC	1.7×10^6	1.2×10^6
BC1%	1.2×10^6	5.5×10^5
BC5%	1.0×10^6	6.0×10^5
BC10%	1.0×10^6	6.5×10^5
C1%	4.1×10^5	1.5×10^6
C5%	2.3×10^6	1.8×10^6
C10%	2.6×10^6	8.2×10^5

Label: SC: soil control, VC: soil + vinasse, BC1%: soil + biofertilizer 1%, BC5%: soil + biofertilizer 5%, BC10%: soil + biofertilizer 10% C1%: soil + biofertilizer 1% + vinasse, C5%: soil + biofertilizer 5% + vinasse, C10%: 10% soil + biofertilizer + vinasse.

Quantification of soil microorganisms was assessed by colony forming units per gram of soil (CFU g^{-1}), using pour plate technique with Plate Counter Agar (PCA) to quantify heterotrophic bacteria and Sabouraud agar to quantify fungi, as set forth in Standard Methods for the Examination of Water and Wastewater (APHA, 1998).

The quantification of heterotrophic soil bacteria and fungi were conducted to verify the existence of microbial consumers of organic matter, which allow the biodegradation process.

The results showed that addition of biofertilizer changed microbial community in soil, as shown in Table 1.

Microbiological analysis indicates a significant presence of heterotrophic bacteria and fungi, probably adapted to consumption of organic matter contained in the residue, which reinforces the effluent biodegradation potential in soil.

Table 1 shows an increased amount of fungi with addition of stillage and an overall gain in all microorganisms with addition of biofertilizer, demonstrating that biofertilizer contributed to the bioaugmentation of soil used in the experiment. This is due to biological and nutritional potential of the biofertilizer, which provides large number of microorganisms and sufficient nutrients for growth of intrinsic soil microbes. Pešaković (2013) concluded that biofertilizer had the highest amount of microorganism in soil, increasing from 10.12 ± 2.40 to 32.83 ± 7.86, and the best yield for strawberry plant when compared to the use of fertilizers. Rawia et al., (2009) found that the use of biofertilizer resulted in an increase in microbial biomass and increased seedling height of wallflower (*Matthiola incana*).

Final Considerations

The biofertilizer affects the quality and soil productivity as well as being a viable alternative for control of pests and diseases in plants, since biofertilizers do not act through a single mode of action or have specificity to control pathogens.

Biofertilizers operate in bioaugmentation and biostimulation of soil, hence increasing its biological biomass; promoting greater absorption of nutrients by plants; improving the physical structure of soil; and accelerating the bioremediation of organic matter contained therein.

Therefore, the use of biofertilizers is a feasible and sustainable alternative, especially for organic farming. Biofertilizers can be produced by farmers themselves, thus reducing costs on imported products as well as producing healthier foods. Biofertilizers are, therefore, extremely important agro-ecological management tools in productive properties.

References

Alexander, M., (1999). Biodegradation and Bioremediation. 2nd ed. San Diego, California, *Academic Press*, p. 453.

Alves, G. S., Santos D., Silva, J. A., Nascimento, J. A. M., Cavalcante, L. F., Dantas, T. A. G., (2009). Estado nutricional do pimentão cultivado em solo tratado com diferentes tipos de biofertilizantes. *Acta Scientiarum. Agronomy*, 31, 661-665.

APHA. (1998). Standard methods for the examination of water and wastewater. 19th ed. AWWA: Washington.

Atlas, R. M., Bartha, R., (1993). Microbial ecology: fundamentals and applications. *Redwood City: The Benjamin/Cummings*, p. 505.

Atlas, R., (1998). Petroleum biodegradation and oil spill bioremediation. *Marine Pollution Bulletin* 31, 178-182.

Baker, K. H. E., Herson, D. S., (1994). Bioremediation. *Environmental Microbiology Associates*, p. 375.

Barbosa, S. A., Medeiros, B. M., (2007). Potencial De Ação Elicitora dos Biofertilizantes Líquidos Na Indução De Resistência Sistêmica Vegetal. *Revista Brasileira de Agroecologia*. 2, 1453-1457.

Bernoth, L., Firth, I., MCallister, P., Rhodes, S., (2000). Biotechnologies for remediation and pollution control in the mining industry. *Minerals and Metallurgical Processing*, 17, 105-111.

Bettiol, W., Stadnik, M. J., (2001). Controle alternativo de oídios. In: Stadnik, M. J. Oídios. Jaguariúna: *Embrapa Meio Ambiente*, 165-192.

Bettiol, W., Tratch, R., Galvão, J. A. H., (1998). Controle de doenças de plantas com biofertilizantes. Jaguariúna: EMATER/ CNPMA. 22p. (EMBRAPA-CNPMA. Circular Técnica, 02).

Ceretta, C. A., Durigon, R., Basso, C. J., Barcellos, L. A. R., Vieira, F. C. B., (2003). Características químicas de solo sob aplicação de esterco líquido de suínos em compostagem natural. *Pesquisa Agropecuária Brasileira*, 38, 729–735.

Cutright, T. J., Lee, S., (1994). In Situ Soil Remediation: Bacteria or Fungi? Department of Chemical Engineering, University of Akron. Akron - Ohio - USA. s/v, 413-419.

D'Andrea, P. A., (2002). Processo de compostagem líquida contínua- CLC e Biofertilizante. Microbiol Indústria e Comércio LTDA. (BR/SP). N° 2099 Patent Registro PI0207342-0A2.

D'Andrea, P.A., (2010). Processo de compostagem líquida contínua – CLC e biofertilizante. BRN. PI0207342-0. A2. Disponível em < www.microgeo.com.br>, acesso em 10 de janeiro de 2015.

Damatto Júnior, E. R., Villas Bôas, R. L., Leonel, S., Fernandes, D. M., (2006). Alterações em propriedades de solo adubado com doses de composto orgânico sob cultivo de bananeira. *Revista Brasileira de Fruticultura* 28, 546-549.

Deleito, C. S. R., Carmo, M. G. F. do, Fernandes, M. C. A., Aboud, A. C.S., (2005). Ação bacteriostática do biofertilizante Agrobio in vitro. *Horticultura Brasileira*, 23, 281-284.

Delgado, A., Madrid, A., Kassem, S., Andreu, L., Campillo, M.C., (2002). Phosphorus fertilizer recovery from calcareous soils amended with humic and fulvic acids. *Plant and soil*, 245, 277-286.

EPA: Environmental Protection Agency, United States. (2004a). How to Evaluate Alternative Cleanup Technologies for Underground Storage Tank Sites: A Guide for Corrective Action Plan Reviewers. Chapter XII - Enhanced Aerobic Bioremediation, EPA 510-B-94-003, EPA 510-B-95-007 and EPA 510-R-04-002, 73p.

EPA: Environmental Protection Agency, United States. (2004b). How to Evaluate Alternative Cleanup Technologies for Underground Storage Tank Sites: A Guide for Corrective Action Plan Reviewers Chapter IV - Biopiles, EPA 510-B-94-003, EPA 510-B-95-007 and EPA 510-R-04002, 30p.

Freire, W. J., Cortez, L. A. B., (2000). Vinhaça de cana-de-açúcar. Guaíba: *Agropecuária*, p. 203.

Furlanetto, C., Café Filho, A.C., Tomita, C.K., Cavalcanti, M.H., (1996). Eficiência de fungicidas e de um composto líquido bioativo no controle da flor preta do morangueiro no Distrito Federal. *Fitopatologia Brasileira*, 21, p. 399.

Galbiatti, J. A., Garcia, A., Silva, M. L., Mastrocola, M. A., Caldeira, D. S. A., (1996). Efeitos de diferentes doses e épocas de aplicação de efluente de biodigestor e da adubação mineral em feijoeiro-comum (*Phaseolus vulgaris L.*) submetido a duas lâminas de água por meio de irrigação por sulco. *Científica*, Jaboticabal, 24, 63-74.

Hammer, M. J., Hammer Júnior, M. J., (2004). Water and wastewater tecnology. 5th.ed. New Jersey: *Prentice Hall International*, 540.

Huesemann, M. H., (1994). Guidelines for Land-Treating Petroleum Hydrocarbon-Contaminated Soils. *Journal of Contamination*, 3, 299-318.

Junqueira, C. Á. R., Molina Júnior, V. E., Lossardo, L. F., Felicio, B. C., Moreira Júnior, O., Foschini, R. C., Mendes, R. M., Lorandi, R., (2009). Identificação do potencial de contaminação de aquíferos livres por vinhaça na bacia do Ribeirão do Pântano, Descalvado (SP), Brasil. *Revista Brasileira de Geociências,* 39, 507-518.

Kozen, E. A., Alvarenga, R. C., (2005). Manejo e utilização de dejetos animais: aspectos agronômicos e ambientais. Sete Lagoas – MG. Empresa Brasileira de Pesquisa Agropecuária- EMBRAPA, *Circular Técnica 63*, 16p.

Machado, A. T., Machado, C. T. T., (2009). Manejo da diversidade genética do milho em sistemas agroecológicos. *EMBRAPA Cerrados*. Ministério da Agricultura, Pecuária e Abastecimento. Planaltina-DF.

Medeiros, M. B., Wanderley, P. A., Wanderley, M. J. A., (2003). Biofertilizantes líquidos: Processo trofobiótico para proteção de plantas em cultivos orgânicos. *Revista Biotecnologia Ciência e Desenvolvimento*, 31, 38-44.

Millioli, V. S., Sobral, L. G. S., Sérvulo, E. F. C., Carvalho, D. D., (2008). Biorremediação de solo impactado com óleo cru: avaliação da potencialidade da utilização surfatantes. *Série Tecnologia Ambiental.* Rio de Janeiro: CETEM/MCT.

Paul, D., Pandey, G., Pandey, J., Jain, K. R., (2005). Accessing microbial diversity for bioremediation and environmental restoration. *Trends in Biotechnology*, 23, 135–114.

Penteado, S, R., (2004). Fruticultura Orgânica: formação e condução. *Viçosa: Aprenda Fácil*, -p. 308.

Pešaković, M., Karaklaji´c-Staji´c, Z., Milenkovi´c, S., Mitrovi´c, O., (2013). Biofertilizer affecting yield related characteristics of strawberry (Fragaria x ananassa Duch.) and soil microorganisms. *Scientia Horticulturae,* 238–243.

Prince, R. C., (1993). Petroleum spill bioremediation in marine environments, *Critical Reviews in Microbiology*, v.19, p.217-242.

Queiróz, F. M., Matos, A. T., Pereira, O. G., Oliveira, R. A., (2004). Características químicas do solo submetido ao tratamento com esterco líquido de suínos e cultivado com gramíneas forrageiras. *Ciência Rural*, Santa Maria, 34, 1487– 492.

Rawia, E. A., Nemat, A. M., Hamouda, H. A., (2009). Evaluate effectiveness of bio and mineral fertilization on the on the growth parameters and marketable cut flowers of Matthiola incana L. American-Eurasian. *Journal of Agriculture and Environmental Sciences*. 5, 509–518.

Rhoades, J. D., Kandiah, A., Mashali, A. M., (2000). Uso de águas salinas para produção agrícola. Campina Grande: UFPB, 117p. Estudos FAO Irrigação e drenagem, 48p.

Ribeiro, G. L., Lopes, J. C., Martins Filho, S., Ramalho, S. S., (2000). Adubação orgânica na produção do pimentão. *Horticultura Brasileira*, 18, 134-137.

Santos, A. C. V., (1991). Efeitos nutricionais e fitossanitários do biofertilizante liquido a nível de campo. *Revista Brasileira de Fruticultura*, 13, 275–279.

Santos, A. C. V., (1992). Biofertilizantes líquidos: o defensivo agrícola da natureza. 2 ed., *rev. Niterói*: EMATER – RIO, p. 162.

Sediyama, M. A. N., Santos, M. R., Vidigal, S. M., Salgado, L. T., Pedrosa, M. W., Jacob, L. L., (2009). Produtividade e estado nutricional do quiabeiro em função da densidade populacional e do biofertilizante suíno. *Bragantia*, 68, 913-920.

Silva, A. J. N., Ribeiro, M. R., (1998). Caracterização de um Latossolo Amarelo sob cultivo contínuo de cana-de-açúcar no Estado de Alagoas: propriedades químicas. *Revista Brasileira de Ciência do Solo*, v.22, n.2, p.291-299.

Silva, E., Fialho, A. M., Sa-Correia, I., Burns, R. G., Shaw, L. J., (2004). Combined bioaugmentation and biostimulation to cleanup soil contaminated with high concentrations of atrazine. *Environmental Science & Technology,* 38:632-637.

Sims, J. L., Suflita, J. M., Russell, H. H., (1992). Ground water issue: In situ bioremediation of contaminated ground water, EPA/540/S-92/003, U.S. Environmental Protection Agency Office of Solid Waste and Emergency Response, 11p.

Souza, J. L., Resende, P., (2003). *Manual de Horticultura orgânica.* Viçosa: Aprenda Fácil. p.564.

Tanaka M. T., Sengik E., Santos H. S., Habel J. C., Scapim, C. A., Silvério, L., Kvitschal, M. V., Arquez, I. C., (2003). Efeito da aplicação foliar de biofertilizantes, bioestimulantes e micronutrientes na cultura do tomateiro (*Lycopersicon esculetum* Mill.) *Acta Scientiarum. Agronomy*, 25, 315-321.

USEPA, Environmental Protection Authority., (2005). *Guidelines Soil bioremediation.* U.S. Environmental Protection Agency,Washington, DC, 1–9.

Vessey, J. K., (2003). Plant growth promoting rhizobacteria as biofertilizers. *Plant and Soil,* 255, 571–586.

Vidali, M., (2001). Bioremediation: an overview. *Pure and Applied Chemistry*, 73, 1163–1172.

Vidigal, S. M., Sediyama, M. A. N., Pedrosa, M. W., Santos, M. R., (2010). Produtividade de cebola em cultivo orgânico utilizando composto à base de dejetos de suínos. *Horticultura Brasileira.* 28: 168-173.

In: Soil Management
Editor: Manuel Esteban Lucas Borja
ISBN: 978-1-63483-274-8

Chapter 8

ADVANCES ON REMOTE SENSING AS SUPPORTING TOOL FOR HYDROLOGICAL MANAGEMENT

***A. L. Vázquez*[1]*, M. J. Oates*[1]*, A. Ruiz-Canales*[1]*, J. M. Molina-Martínez*[2]*, D. G. Fernández-Pacheco*[2]*, L. Ruiz-Peñalver*[2] *and V. A. Rodrigues*[3]**

[1]Universidad Miguel Hernández de Elche, Spain
[2]Universidad Politécnica de Cartagena, Spain
[3]Universidade Estadual Paulista "Julio de Mesquita Filho", Brazil

ABSTRACT

Over the last decades remote sensing has become a very valuable tool for hydrology, especially in water management. This development is due to, on the one hand, the fast technological evolution registered on sensors and platforms, thus enabling resolutions (spatial, temporal, radiometric...) ever even thought about before and every day at a lower cost; on the other hand, methodological research has brilliantly followed technical advances, so that in a blink we have passed from eye crop identification to Geographic Information Systems and sophisticated physic-physiologic computer models. Nonetheless, Remote Sensing is not too much applied yet in the operational level; thence, there is still very much to do in order to improve user-friendliness, reliability and accuracy at every level, from the field point to the continent.

Keywords: hydrology, classification, evapotranspiration, groundwater, mapping, salinity, water management

Introduction

Quantitative assessment of the soil-vegetation-atmosphere transfer processes can lead to a better understanding of the relationships between forestry growth and water management.

As freshwater becomes an increasingly scarce resource, all opportunities to better manage water uses, particularly in hydrology must be taken. An important challenge in the field of water resources is to utilize the timely, objective and accurate information provided by remote sensing (RS hereafter). However, even though considerable progress has been made over the past 30 years in research applications, remotely sensed data remain underutilized by practicing water resource managers.

The application of RS techniques provides several advantages over field measurements. First, measurements derived from remote sensing are objective; they are not based on opinions. Second, the information is collected in a systematic way which allows time series and comparison between schemes. Third, remote sensing covers a wide area such as entire river basins. Fourth, information can be aggregated to give a bulk representation, or disaggregated to very fine scales to provide more detailed and explanatory information related to spatial uniformity.

And fifth, RS as a tool for dynamic data capture, combined with tables (data from wells, piezometers, river flows...) and digital maps (topographic maps, geological maps, cadastral data at a parcel scale,...) in a Geographic Information System (hereafter, GIS), allows the generation of accurate information at a reasonable cost (Castaño et al., 2001).

A tentative classification of the potential uses of RS on water management (Martín de Santaolalla and Calera, 2005) is adapted to guide the structure of the present chapter: inundation and forestry zones identification; thematic mapping; detection of salinity problems; water management and assessment.

Inundation and Forestry Zones Identification: Thematic Mapping

Thematic mapping is one of the main goals of remote sensing because of their multiple applications (Martín de Santaolalla and Calera, 2005). These maps can be the final work output or just an intermediate step to gather information about land use and vegetation cover.

The main features supporting the use of remote sensing in this field are (Montesinos et al., 2007): i) objectivity: the data is composed by digital imaging, ii) continuity: the information is not an extrapolation from punctual measurements, but a discrete representation of the continuum (pixels), generating a digital image, iii) temporal resolution: satellites "visit" every site each certain fixed number of days, iv) other platforms, like Unmanned Aerial Vehicles (UAV), allow the user to choose the observation timing, v) multispectrality, vi) capacity to work in different scales.

Identification of forest is not self-evident for humid climates with regular showers and frequently overcast skies, although successes have been reported by using microwave data to obtain brightness temperature under any weather conditions (Schmugge et al., 2002).

The diverse water management methods can be evaluated and compared through a parametric evaluation system which is primarily based on physical and chemical soil properties (Albaji et al., 2010).

A capability index (Table 1) is calculated for every water management method and registered on a land suitability map (Figure 1) according with: physical soil properties (texture, permeability and available water content); chemical properties interfering with the calcium carbonates status and the salinity/alkalinity status (soluble salts, exchangeable Na, or EC; drainage properties (depth of ground water); environmental factors (slope).

Table 1. Suitability classes for the irrigation capability indices (Ci) classes

Capability Index	Definition	Symbol
>80	Highly suitable	S1
60-80	Moderately suitable	S2
45-59	Marginally suitable	S3
30-44	Currently not-suitable	N1
<29	Permanently not-suitable	N2

Albaji et al., 2010.

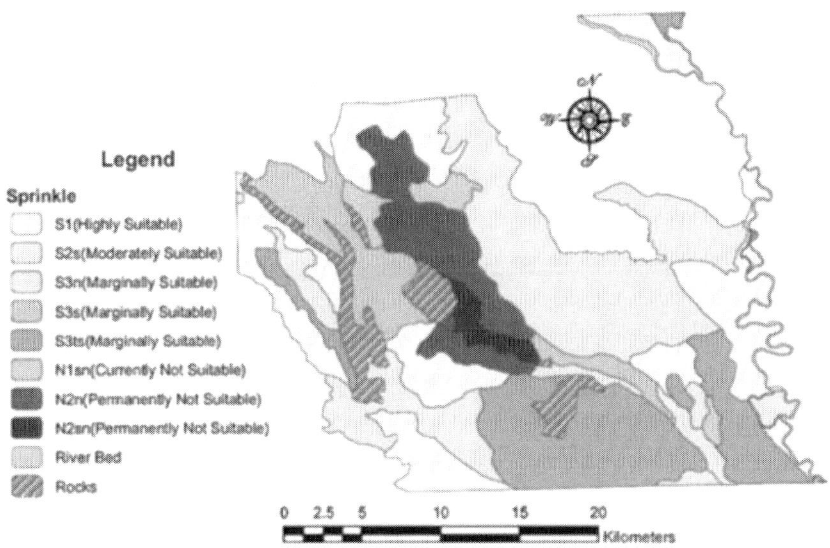

Albaji et al., 2010.

Figure 1. Land suitability map for sprinkler irrigation.

Among data obtained from RS, growth phenology information can be extracted and later applied to classify Land Use and Land Cover (LULC hereafter) (Cheema and Bastiaanssen, 2010).

LULC is one of the backbones of RS since it is required for modeling of the water balance and preparation of water accounting. LULC can be obtained by integrating NDVI (Normalized Difference Vegetation Index) time series with ground information and expert knowledge on the growing patterns (phenology) of the forest species.

The discrimination procedure uses the various spectral responses over a period of time according to the vegetal specie's phenological evolution. Temporal monitoring of vegetal forest species evolution during the growing season is achieved by means of NDVI from scenes obtained on the most suitable dates according to the phenological evolution of the species involved. An example of a land use and land cover map developed from NDVI values covering a vegetation phenological cycle is illustrated in Figure 2.

A large supply of water under semiarid conditions entails a better growth of the vegetation and, as a result, also a greater degree of land cover, magnitude correlating directly with the NDVI index.

Multitemporal supervised classification is based on imagery obtained on different dates from different platforms (Landsat, IRS, SPOT, ...).

The use of these varied data requires: (a) accurate georeferencing to ensure proper overlaying of images and (b) multisensor NDVI which must be obtained from calibrated reflectances. The goal is to develop an up to date, regionally consistent and detailed LULC map.

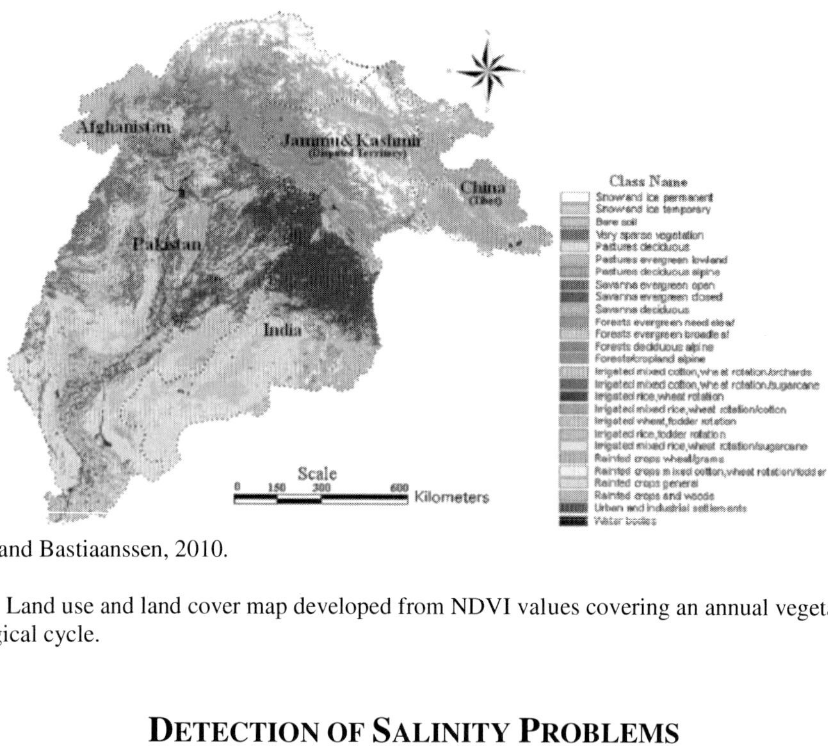

Cheema and Bastiaanssen, 2010.

Figure 2. Land use and land cover map developed from NDVI values covering an annual vegetation phenological cycle.

DETECTION OF SALINITY PROBLEMS

Salinization and soil degradation occurs in areas where saline groundwaters rise towards the ground surface and evaporation exceeds precipitation (Dehaan and Taylor, 2002).

In these zones, capillary rise of saline groundwater causes the direct precipitation of evaporate minerals at the surface. Salinization may also occur when salts concentrate in soils by the evaporation of freestanding irrigation water.

Ground-based electromagnetic (EM) measurements of soil electrical conductivity are generally accepted as the most effective proximal sensing method for rapid acquisition and quantification of soil salinization. Despite this, efforts are being made to develop more cost-effective methods of mapping soil salinization.

These methods include airborne EM techniques, airborne radar, and hyperspectral remote sensing techniques in the 400 to 2500 nm spectral range (bare soil or vegetation).

Additional field spectrometry, calibrations and visits to the field are required. One sample of this system can be seen in Figure 3, where a synthetic true color composite of HyMap imagery using Bands 3, 9, and 15 (Figure 3a) and a distribution of three salinized land terrain element indicators: (highly salinized soil, samphire, and sea blite) (Figure 3b) can be appreciated.

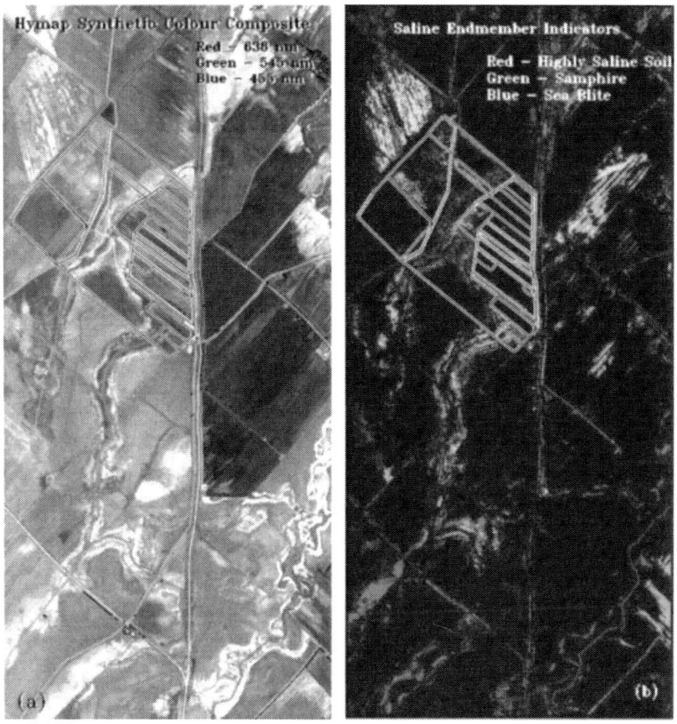

Dehaan and Taylor, 2002.

Figure 3. (a) Synthetic true color composite of HyMap imagery. (b) Distribution of three salinized land terrain element indicators: highly salinized soil (red), samphire (green), and sea blite (blue).

Compared with normal soils, the highly salinized ones show additional distinct absorption features at 680, 1180 and 1780 nm, a reduction of the 2200 nm absorption feature intensity (loss of crystallinity in the clay minerals), a high reflectance at 800 nm, and a decrease in the overall slope of the curve between 800 and 1300 nm as samples become more salinized, all accompanied by a change in soil color from red to grey. Ground observations and radiometric measurements show that quantity and mineralogy of salts are the main factors affecting salt reflectance. Salty crusts use to be smoother than non-saline surfaces and have an overall higher reflectance in the visible (especially blue) and near-infrared wave bands (Metternicht and Zinck, 2003).

Notwithstanding this, the tendency of salts to concentrate locally, in patchy spots, creates variable mixtures of features within individual pixels. In addition, saline crusts vary considerably in structure, from smooth to rough, and in color, from white to dark, causing large variations in reflectance. Direct and precise estimation of salt quantities from remote sensing is rather difficult because of: (a) lack of specific absorption bands of some salt types (e.g., halite) and occurrence of spectral confusions, (b) low spectral resolution of certain sensors, (c) vertical, spatial, and temporal variability of salinity in the soil mantle and substratum, (d) the presence of other soil chromophores (e.g., structural crusts, soil surface texture, soil color, soil surface roughness as determined by different salt types). Ground validation is mandatory. Detection and mapping of salt affected areas can be implemented with several different sensors, which are summarized in Table 2.

While EM methods measure soil conductivity properties, optical remote sensing methods measure the spectral properties of surface materials. Therefore, a high level of agreement between "salinization" maps produced by these two methods is unlikely since each method measures the effects of salinization in different ways.

In passive (e.g., optical and infrared) RS data acquisition, it must be taken into account that salt identification is easiest at the end of the dry season. On the other hand, the characterization of saline soils using active (radar) imagery requires some soil moisture, as the measurements depend principally on dielectric constant. Detection accuracy can be improved by combining field data and multitemporal or multispectral image data, supported by GIS technology. An integrated approach to assess soil salinity encompasses spatial analysis of ground truth and satellite or aerial data (Abbas and Khan, 2006). Out of several salinity indices based on data measured remotely by the Indian sensor IRS1-B LISS-II, the following one happens to be very well related to ground truth data:

Table 2. Ground-based air- and satellite borne sensors reported in the literature for detection and mapping of salt-affected areas

Sensor	No. of bands	Spectral range (μm)	Spatial Resolution
Satellite-borne Sensors			
Landsat TM4, TM5 TM7-ETM+	7 (1-7), 8 (1-8)	Visible, NIR, MIR, TIR B1: 0.45–0.52 B2: 0.52–0.60 B3: 0.63–0.69 B4: 0.76–0.90 B5: 1.55–1.75 B6: 10.40–12.50 B7: 2.08–2.35 B8: 0.52–0.90 (pan)	Bands 1-5 and 7: 30 m. Band 6: 120 m Band 8: 15 m
SPOT 1 – 3	4 (Xs1–3 and Pan)	Visible, NIR Xs1: 0.50–0.59 Xs2: 0.61–0.68 Xs3: 0.79–0.89	Xs and Xi: 20 m,Pan and Mono: 10 m
SPOT 4	5 (Xi1–4 and Mono)	Xi4: 1.58–1.75 Pan: 0.51–0.73 Mono: 0.61–0.68	
LISS-III	4	Visible, NIR, midinfrared B1: 0.52–0.59 B2: 0.62–0.68 B3: 0.77–0.86 B4: 1.55–1.70	Bands 1 – 3: 23 m, Band 4: 70 m
LISS-II	4	Visible, NIR B1: 0.45–0.52 B2: 0.52–0.59 B3: 0.62–0.68 B4: 0.77–0.86	36.25 m
IRS-1C	1	0,5-0,75 (PAN)	5.8 m
JERS-1	1	Microwave L-band (23 cm)-HH pol	18-12.5 m
Airborne sensors			
Aerial photographs		B/W; colour infrared	variable, depending on flight height
Narrow-band videography	3	Visible, NIR 0.54–0.55 0.64–0.65 0.84–0.85	3.4 m
DMSV (digital multispectral video systems)	4	Visible, NIR 0.44–0.46 0.54–0.56 0.64–0.66 0.74–0.76	variable: 0.25–2 m
AIRSAR–TOPSAR	3	Microwave P-, L-bands (full polarimetric), C-band (VV polarization)	10 m
Hyperspectral Hymap	128	Visible, NIR, midinfrared 0.45–2.5	2 – 10 m
Hyperspectral DAIS-7915	79	0.4–12 0.4–1 (32 bands) 1.5–1.8 (8 bands) 2–2.5 (32 bands) 3–5 (1 band) 8.5–12.3 (6 bands)	3 – 20 m
Airborne geophysics		Gravity Magnetic Electromagnetic Gamma ray	
Ground-based sensors			
Electromagnetic induction meter (EM31, EM34-3, EM38, EM39)		Electromagnetic conductivity meter measures the bulk electrical conductivity of soils	
Crop Scan multiband radiometer (Skye Instruments, UK)	8	Visible, NIR 0.36–1.1	

Metternicht and Zinck, 2003.

$$\text{Salinity Index} = (B_2 x B_3)/B_1 \tag{1}$$

where B1, B2 and B3 are bands of the satellite IRS-1B LISS-II, in spectral range of band 1 (0.45-0.52 μm), band 2 (0.52-0.59 μm) and band 3 (0.62-0.68 μm) respectively.

The resulting images can be categorized into, for instance, the four distinct classes of saline soils shown in Table 3. The presence of salts at the terrain surface can be detected indirectly as well through vegetation type and growth, as these are affected by salinity. Salinization affects the leaf angle orientation (leaf roll) and increases chlorosis in plants, which is best observed in the near and middle infrared bands. High correlation was found between NDVI values and soil electrical conductivity, allowing to separate saline–alkaline soils from non-affected ones, but not to discriminate between saline and alkaline soils (Wiegand et al., 1996).

WATER MANAGEMENT AND ASSESSMENT

Water Stress

Detection of crop water stress is crucial for efficient irrigation water management and scheduling, especially in semi-arid regions (Boulet et al., 2007). Water stress or second-stage ET corresponds to the reduction in ET due to the limited availability of root zone soil moisture. Water stress results in a drop of actual ET below the potential rate. Its intensity is usually represented by a Stress Factor (S), which is defined as the complement to one of the actual (λET) to potential (λETp) evapotranspiration ratio:

$$S = 1 - \frac{\lambda ET}{\lambda ETp} \tag{2}$$

This factor is equal to zero for energy-limited evapotranspiration (unstressed conditions), and increases towards one for soil-controlled evapotranspiration (water stress conditions).

Hence, water stress can be assessed by measuring evapotranspiration rates. Data in the thermal infrared (TIR) spectrum is linked to soil moisture and thus to the evaporation flux since the surface temperature is obtained through solving the surface energy balance equation. The use of TIR data to monitor stress can be classified into three broad categories (see Table 4).

For each one of these categories (described below), TIR data can be combined or not to a surface energy balance model in order to provide more elaborated information:

1. Using TIR data for an assessment of instantaneous flux patterns:
   a. Combination of TIR radiance and vegetation indices: Crop Water Stress Index (CWSI), Surface Energy Balance Index (SEBI) and Water Deficit Index (WDI), among others. The scaling between both extremes of the stress factor coefficient (S) depends on a second remote sensing variable representing the vegetation cover fraction, which is usually the Normalized Difference Vegetation Index (NDVI), or the Soil Adjusted Vegetation Index (SAVI). This leads to the

classical triangle or trapezoidal shape of the temperature/vegetation cover diagram.

   b. Surface energy balance: SEBAL, SEBS and ALARM. In these cases, surface temperature is used as input to derive the sensible heat flux and obtain λET as a residual of the energy balance.

2. Using time series of TIR observations: CWSI and TDVI indices. These methods use change detection algorithms and are based on the assumption that when the surface enters the 'soil- controlled' stage, there are strong and sudden changes in surface conditions that impact on the surface temperature.
3. Assimilation of TIR data into land surface models. Assimilation 'sensu lato' enables to adjust either a state variable or a parameter of a given state-space process model in order to reduce the difference between the simulated and the observed radiometric surface temperatures.

For each one of these categories (described below), TIR data can be combined or not to a surface energy balance model in order to provide more elaborated information:

1. Using TIR data for an assessment of instantaneous flux patterns:
   a. Combination of TIR radiance and vegetation indices: Crop Water Stress Index (CWSI), Surface Energy Balance Index (SEBI) and Water Deficit Index (WDI), among others. The scaling between both extremes of the stress factor coefficient (S) depends on a second remote sensing variable representing the vegetation cover fraction, which is usually the Normalized Difference Vegetation Index (NDVI), or the Soil Adjusted Vegetation Index (SAVI). This leads to the classical triangle or trapezoidal shape of the temperature/vegetation cover diagram.
   b. Surface energy balance: SEBAL, SEBS and ALARM. In these cases, surface temperature is used as input to derive the sensible heat flux and obtain λET as a residual of the energy balance.
2. Using time series of TIR observations: CWSI and TDVI indices. These methods use change detection algorithms and are based on the assumption that when the surface enters the 'soil- controlled' stage, there are strong and sudden changes in surface conditions that impact on the surface temperature.
3. Assimilation of TIR data into land surface models. Assimilation 'sensu lato' enables to adjust either a state variable or a parameter of a given state-space process model in order to reduce the difference between the simulated and the observed radiometric surface temperatures.

Data in the thermal infrared spectrum is still nowadays the most promising data source to monitor water stress at most scales ranging from the paddock to the region. In addition, the classical Crop Water Stress Index (CWSI) is considered one of the most successful indices, since takes into account the ambient meteorological conditions.

This index is related to (1 - AET/MET), where AET is the actual evapotranspiration and MET is the maximum evapotranspiration, and ranges from 0 for a well-watered crop transpiring at the maximum rate, to 1 for a high water-stressed, non-irrigated crop. CWSI expresses the Stress Factor (S) but it is not valid for all surface conditions because it is restricted to full-canopy conditions and has to be calculated under clear sky conditions.

Instead of this, the empirical approach (CWSIe) is largely used because it only requires three variables to be applied (vapor pressure deficit, and air and canopy temperatures), providing interesting results when applied in arid or semi-arid climates. On the contrary, the theoretical approach (CWSIt), based on energy balance considerations, was often stated to be more adapted for humid regions. However, it requires several variables which are not always available, thus this approach is discarded for operational irrigation scheduling.

More recent methods to study water stress with the help of thermal infrared data have shifted away from the stress factor concept to more complex instantaneous diagrams based on surface cover or complex mathematical methods such as data assimilation.

Table 4. Classification of some methods to assess water stress from TIR data

Classified category	With energy balance model	Without energy balance model
Instantaneous retrieval	SEBAL, ALARM, SEBS, etc.	TVI, TVDI, etc.
Times series analysis	CWSI, SEBI, WDI, etc.	Albedo, T_s -T_a, (T_s - T_a)/R_s, etc.
Data assimilation in a state-space model	Simple (thermal inertia only, no water balance involved) or complex (SVAT, water balance involved)	Assimilation of time to stress in water balance models (e.g., SVATsimple)

Boulet et al., 2007.

Besides CWSI, the so called stomatal conductance index (Ig) is an indicator specifically developed for infrared thermography, so that it can be used to estimate water stress (Maes et al., 2011). The Ig term is calculated from the measured leaf temperature and the temperatures of a wet and dry reference surface following the equation:

$$I_g = \frac{T_{dry} - T_l}{T_l - T_{wet}} = \frac{g_S}{G} \quad (3)$$

where gS is the stomatal conductance and G is a parameter independent of net radiation and vapor pressure deficit. Unfortunately, Ig coefficient shows serious shortcomings on its application.

Advances in sensor miniaturization have allowed to estimate stomatal conductance and CWSI from high resolution (better than 0,5 m) thermal imaging (TIR) retrieved from a compact TIR sensor on board of a micro UAV (Unmanned Aerial Vehicle) (Berni et al., 2009).

The use of high-resolution remote thermal imagery (Airborne Hyperspectral Scanner, Advanced Spaceborne Thermal Emission and Reflection radiometer, and cutting edge UAV sensing) enables to identify individual trees under different transpiration levels. An image detail showing the spatial resolution differences of AHS (2 m) against the UAV (40 cm) can be seen in Figure 4, where the spatial resolution of the UAV imagery shows individual tree crown, enabling pure crown temperature extraction.

The Discrete Anisotropic Radiative Transfer (DART) 3D model simulates tree orchards, separating pure vegetation, soil and shadows, so it is possible to discriminate between irrigated and rainfed open-tree canopies (Sepulcre-Canto et al., 2009). These results are particularly important since ET estimation models (like SEBAL or METRIC) models use NDVI and surface temperature as inputs, potentially leading to erroneous ET estimations.

Scheduling

Under Precision Farming, Remote Sensing and Geographic Information System techniques have played a vital role on defining variable rate applications according to various management zones requirements. There has been an increasing awareness that in many situations it is best to vary the amount of water applied spatially (Ruiz-Canales et al., 2010; Melián et al., 2010), due to, for example: (i) variable crop demands, (ii) soil variability, and (iii) non-irrigation zones.

Hence, Variable Rate Application (VRA) (water, fertilizer, chemicals, and so on) is now considered the backbone of Precision Farming. Variable Rate Technology (VRT) allows variable irrigation scheduling and tillage throughout any irrigation system.

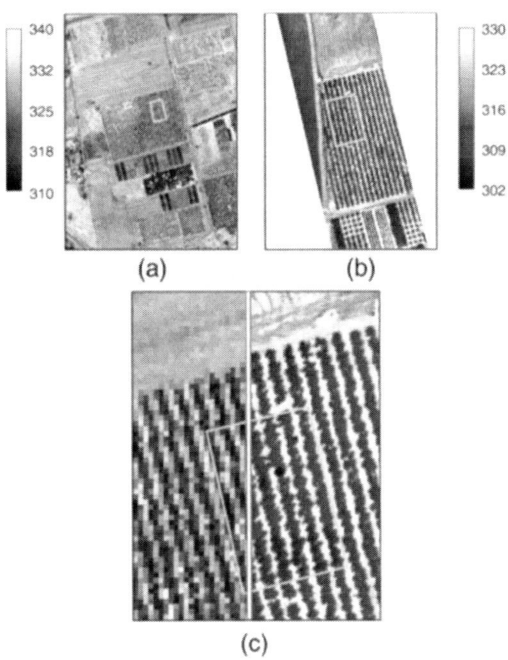

Berni et al., 2009.

Figure 4. Airborne thermal imagery acquired over the study site: a) AHS image; b) UAV image; c) image detail showing the spatial resolution differences.

The irrigation rate can change through information gathered by satellite or aerial sensors in conjunction with thematic mapping, GPS and GIS. The VRA of water is based on the surface energy balance algorithm for land (SEBAL) model, the Penman-Monteith model and RS, being used altogether for estimating crop evapotranspiration. The process is as follows:

Retrieving land surface temperature (LST), by using thermal band of i.e., ETM+. Pre-processing parameters required for SEBAL: NDVI, emissivity, albedo, surface temperature, net incoming radiation, soil heat flux, etc. → ET calculation.

Calculation of crop water requirements and irrigation scheme based on soil, climate and crop data: CropWat model, based on FAO Penman-Monteith approach.

Existing irrigation decision support systems include the modeling approach (regional or site-specific), direct crop stress measurement, soil moisture measurements (Figure 5), or a combination of these methods. A good irrigation decision support tool would ideally address the spatial and temporal variability of crop demand and soil supply, and should be compatible with recent advances in full automation of irrigation systems (Hedley and Yule, 2009). A fundamental component of any irrigation scheduling is precipitation. Remote Sensing of precipitations is accomplished with the aid of ground based radar and satellite retrievals (Michaelides et al., 2009). Rainfall intensity retrievals from space-borne sensors have been based mostly on passive sensors for about 30 years using visible (VIS), infrared (IR) and passive microwave (PMW) wavelengths on geostationary (GEO) and low Earth orbiting (LEO) satellites. Since 1997 there is an operational satellite active sensor, the Tropical Rainfall Measuring Mission.

Additionally, the high variability of soil types often complicates the use of traditional irrigation scheduling methods. Some of these methods are based on water balance computation that requires soil measurements that are time consuming and not easy to implement (Abrisqueta et al., 2001; Plana et al., 2002).

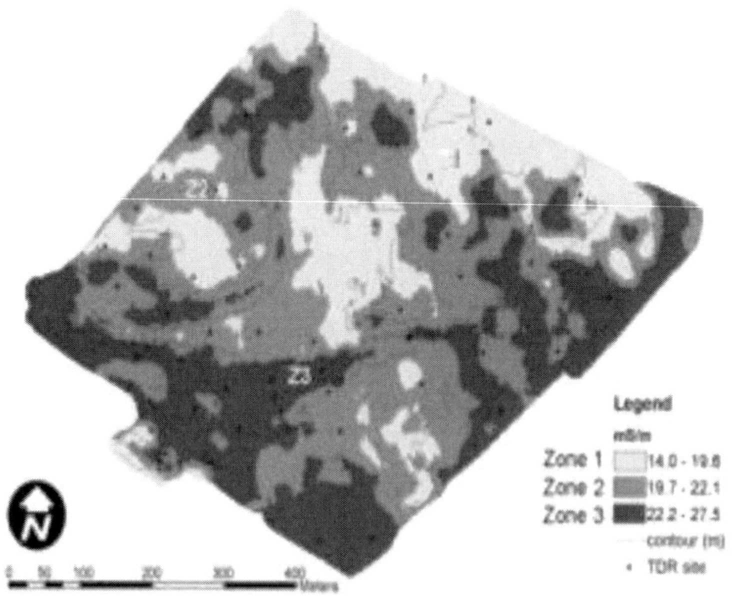

Hedley and Yule, 2009.

Figure 5. Soil apparent electrical conductivity map of the study area (Z1, Z2 and Z3 = Aquaflex positions in Zone 1, Zone 2 and Zone 3 respectively).

The solution can be found in water balance models, as OSIRI, a decision-making tool for monitoring irrigation of small farms in heterogeneous environments. The tool is based on a simple water balance simulation model (PROBE) (Chopart and Vauclin, 1990) coupled with a comprehensive set of decisions rules. The tool simulates the water balance to come and needs local soil and root information and regular updating (actual rainfalls). It is also recommended to check the water status of the crop in situ, being thermal infrared canopy temperature a serious candidate for the task.

Performance Assessment

Performance assessment is an essential component of effective water management. Regular feedback of information from the field into water management decision making can substantially improve the performance of water delivery services. Remote Sensing may now provide viable solutions in some situations, allowing repeated sampling of field conditions in units as small as 100 ha. In order to get the best redistribution, water accounting studies are used to identify the various uses of water in the basin wide context and the productivity of each use. The monitoring of crop production and irrigation at a regional scale can be based on the use of ecosystem process models and RS data. The former simulate the temporal dynamics of the biophysical variables which affect crop photosynthesis and water consumption at a fine time step (hourly or daily); the latter allows providing the spatial distribution of these variables over a region of interest at a time span of 10 days to a month. This explains the growing interest about using remotely-sensed information in ecosystem process models (Duchemin et al., 2006).

Remote sensing based estimations of water consumption and water stress combined with secondary agricultural production data can provide good estimations of irrigation performance, water productivity and the link between both parameters at a variety of scales (Ahmad et al., 2009). Better estimations of crop area and actual water consumption are required, since surface water supplies are not only used directly in the field, but also provide a substantial, but unquantified portion of groundwater recharge. RS estimations of actual ET take account of factors that reduce water consumption below potential, such as salinity, crop condition and poor irrigation scheduling.

Since irrigation agriculture is usually the main water consumer, policy directives are calling for tools to aid operational monitoring in planning, control and charging of irrigation water (Folhes et al., 2009).

Opportunely, satellite images have provided new possibilities to fulfill this quantization need, and now satellite-based models are finally in place for application to agricultural extensive areas, as for instance, the aforementioned METRIC model.

By combining METRIC-derived monthly evapotranspiration information for the irrigable area with monthly average depleted fraction calculated for a number of reference parcels, it is possible to predict the irrigation water applied over the entire irrigation scheme.

This application assumes there is a correlation between METRIC evapotranspiration and total irrigation water applied. However, satellite-based water consuming mapping will never completely substitute ground irrigation water metering, since some measurement data related to individual parcels are needed to capture the pattern of consumption which in turn reflects the irrigation practice on a diverse crop mix area.

One step forward of using SEBAL and METRIC is the introduction of inverse modeling in order to establish the required accuracy and temporal frequency for effective use of those techniques (Droogers et al., 2010). One possible way of inverse modeling ET estimation is based on:

a. SEBAL and METRIC models to obtain daily actual evapotranspiration from satellite multispectral images (Landsat Thematic Mapping images, e.g.,)
b. Application of SWAP (soil-water-atmosphere-plant) model to explore the impact of temporal data resolution on estimating actual irrigation applications. SWAP is an integrated physically based simulation model for water, solute, and heat transport in the saturated–unsaturated zone in relation to crop growth. The SWAP model requires various data as input which can be divided into state variables, boundary conditions (model forcing) and calibration/ validation data. The most important state variables are related to soil and crop characteristics.
c. Model calibration by using the PEST (Parameter ESTimation) program. PEST is a non-linear parameter estimation package that can be used to estimate parameters for just about any existing computer model. PEST runs a model as many times as necessary to adjust its parameters until selected model outputs match a complementary set of field or laboratory measurements as closely as possible.

Conclusion

Remote sensing is considered a multidisciplinary methodology which is in continuous development and that offers more and more possibilities of application in the hydrology and irrigation management domains, reaching to offer all kinds of services at every level, from the horticultural plot to the continent. One reason for the increasing relevancy of the remote sensing techniques is found in the possibility of creating all kind of digital cartographies and combining them with Geographic Information Systems, enabling in this way temporal and spatial hydrological planification, monitoring of water consumptions, estimation and mapping of multiple parameters such as evapotranspiration, soil moisture, groundwater outcrops, water turbidity, salinity problems, and so on.

In this work, the state of the art methodologies utilized in remote sensing is revised, covering from the manual methods to the employment of satellite or aerial sensors. The recent use of Unmanned Aerial Vehicles (UAVs) as an economic and reliable option is also worth noting. In the same manner, sensors are constantly experimenting in terms of resolution and, most importantly, of miniaturization. The utilization of multi/hyper spectral sensors, radars, and thermic cameras is also contemplated.

REFERENCES

Abbas, A., Khan, S., 2006. Using Remote Sensing Techniques for Appraisal of Irrigated Soil Salinity. *International Congress on Modelling and Simulation*. Christchurch, New Zealand.

Abrisqueta, J. M., Ruiz, A., Franco, J. A., 2001. Water balance of apricot trees (Prunus armeniaca L. cv. Bulida) under drip irrigation. *Agr. Water Manage.*, 50(3), 211-227.

Ahmad, M. D., Turral, H., Nazeer, A., 2009. Diagnosing irrigation performance and water productivity through satellite remote sensing and secondary data in a large irrigation system of Pakistan. *Agr. Water Manage.*, 96(4), 551-564.

Albaji, M., Shahnazari, A., Behzad, M., Naseri, A., Boroomandnasab, S., Golabi, M., 2010. Comparison of different irrigation methods based on the parametric evaluation approach in Dosalegh plain: Iran. *Agr. Water Manage.*, 97(7), 1093-1098.

Berni, J. A. J., Zarco-Tejada, P. J., Sepulcre-Cantó, G., Fereres, E., Villalobos, F. J., 2009. Estimación de conductancia estomática y detección de estrés hídrico en vegetación mediante imágenes térmicas de alta resolución espacial obtenidas con un vehículo aéreo no tripulado (UAV). *XIII Congreso de la Asociación Española de Teledetección*. Calatayud. [In Spanish].

Boulet, G., Chehbouni, A., Gentine, P., Duchemin, B., Ezzahar, J., Hadria, R., 2007. Monitoring water stress using time series of observed to unstressed surface temperature difference. *Agr. Forest Meteorol.*, 146(3-4), 159-172.

Castaño, S., Denore, B. J., Montesinos, S., Sánchez, I., Quintanilla, A., 2001. Técnicas de observación de la Tierra en la gestión de recursos hídricos en áreas del mediterráneo. Proyecto Isla. *IX Congreso Nacional de Teledetección*. [In Spanish].

Cheema, M. J. M., Bastiaanssen, W. G. M., 2010. Land use and land cover classification in the irrigated Indus Basin using growth phenology information from satellite data to support water management analysis. *Agr. Water Manage.*, 97(10), 1541-1552.

Chopart, J. L.,Vauclin, M., 1990. Water-balance estimation model-field-test and sensitivity analysis. *Soil Sci. Soc. Am. J.*, 54(5), 1377-1384.

Dehaan, R. L., Taylor, G. R., 2002. Field-derived spectra of salinized soils and vegetation as indicators of irrigation-induced soil salinization. *Remote Sens. Environ.*, 80(3), 406-417.

Droogers, P., Immerzeel, W. W., Lorite, I. J., 2010. Estimating actual irrigation application by remotely sensed evapotranspiration observations. *Agr. Water Manage.*, 97(9), 1351-1359.

Duchemin, B., Hadria, R., Erraki, S., Boulet, G., Maisongrande, P., Chehbouni, A., Escadafal, R., Ezzahar, J., Hoedjes, J. C. B., Kharrou, M. H., Khabba, S., Mougenot, B., Olioso, A., Rodriguez, J. C., Simonneaux, V., 2006. Monitoring wheat phenology and irrigation in Central Morocco: On the use of relationships between evapotranspiration, crops coefficients, leaf area index and remotely-sensed vegetation indices. *Agr. Water Manage.*, 79(1), 1-27.

Folhes, M. T., Renno, C. D., Soares, J. V., 2009. Remote sensing for irrigation water management in the semi-arid Northeast of Brazil. *Agr. Water Manage.*, 96(10), 1398-1408.

Hedley, C. B., Yule, I. J., 2009. A method for spatial prediction of daily soil water status for precise irrigation scheduling. *Agr. Water Manage.*, 96(12), 1737-1745.

Maes, W. H., Achten, W. M. J., Reubens, B., Muys, B., 2011. Monitoring stomatal conductance of Jatropha curcas seedlings under different levels of water shortage with infrared thermography. *Agr. Forest Meteorol.*, 151 (5), 554-564.

Martín de Santaolalla. F. J., Calera. A., 2005. Uso de la Teledetección en el seguimiento de los cultivos de regadío. *Ed. Mundiprensa*, pp. 523-582. [In Spanish].

Melián, A., Molina, J. M., Ruiz-Canales, A., 2010. La automatización y el telecontrol en riego, herramientas para el ahorro de agua: aplicación al cultivo de brócoli en el Sudeste español. *Agrícola vergel: Fruticultura, horticultura, floricultura*, 29(341), 299-306. [In Spanish].

Metternicht, G. I., Zinck, J. A., 2003. Remote sensing of soil salinity: potentials and constraints. *Remote Sens. Environ.*, 85(1), 1-20.

Michaelides, S., Levizzani, V., Anagnostou, E., Bauer, P., Kasparis, T., Lane, J. E., 2009. Precipitation: Measurement, remote sensing, climatology and modeling. *Atmos. Res.*, 94(4), 512-533.

Montesinos, S., Bea, M., Fernández, L., 2007. Estimación de la superficie en regadío mediante técnicas de teledetección: conceptos y su aplicación operativa. *Jornadas sobre las Aguas subterráneas en España ante las Directivas europeas: retos y perspectivas.* Santiago de Compostela. [In Spanish].

Plana, V., Ruiz, A., Ruiz-Sánchez, M. C., Franco, J. A., Abrisqueta, J. M., 2002. Spatial representativity of the possible sites for measuring the water balance of apricot trees. *Agr. Water Manage.*, 57(2), 145-153.

Ruiz-Canales, A., Gonzalez, E., Rodriguez, J. A., Molina, J. M., Melián, A., 2010. Automatización y telecontrol de la gestión del riego para la mejora de la eficiencia energética en proyectos de campos de golf. *Montajes e instalaciones: Revista técnica sobre la construcción e ingeniería de las instalaciones*, 40(450), 42-48. [In Spanish].

Sepulcre-Canto, G., Zarco-Tejada, P. J., Sobrino, J. A., Berni, J. A. J., Jimenez-Munoz, J. C., Gastellu-Etchegorry, J. P., 2009. Discriminating irrigated and rainfed olive orchards with thermal ASTER imagery and DART 3D simulation. *Agr. Forest Meteorol.*, 149(6-7), 962-975.

Schmugge, T. J., Kustas, W. P., Ritchie, J. C., Jackson, T. J., Rango, A., 2002. Remote sensing in hydrology. *Adv. Water Resour.*, 25(8-12), 1367-1385.

Wiegand, C., Anderson, G., Lingle, S., Escobar, D., 1996. Soil salinity effects on crop growth and yield - Illustration of an analysis and mapping methodology for sugarcane. *J. Plant Physiol.*, 148(3-4), 418-424.

In: Soil Management
Editor: Manuel Esteban Lucas Borja
ISBN: 978-1-63483-274-8

Chapter 9

HYDROPHYSICS: WATER DYNAMICS IN SOIL PROFILE

***Valdemir Antonio Rodrigues*[1,*]*, Antonio Ruiz Canales*[2]*, Marcos Antonio Rezende*[3] *and Vladimir Eliodoro Costa*[3]**

[1]Departament of Forest, Faculty of Agricultural Science, University Estadual Paulista, FCA/UNESP, Botucatu, São Paulo, Brazil
[2]Escuela Politécnica Superior de Orihuela, Universidad Miguel Hernández de Elche, Orihuele, Spain
[3]Institute of Bioscience, IBB / UNESP, Botucatu,São Paulo, Brazil

ABSTRACT

On its different stages of growth, the vegetation shows the dynamic of the consumption of water from the soil by evapotranspiration. The remaining water is stored according to soil texture and, as time goes by, it reaches the groundwater table. The objective of this work is to study the relation between humidity, density, porosity and shrinkage of the floodplain soil and riparian vegetation and their ability to store water. For this purpose, two locations for every type of soils were evaluated.

Both were placed at the Agronomy University (Faculdade de Ciências Agronômicas) in São Manuel, State of São Paulo, Brazil. The riparian stands soil had native vegetation, the floodplain soil was vegetated with Southern Cattail (Typha domingensis). In both places, soil samples were collected from several depths: 0, 30, 60 and 100 cm. Results show that lower soil density values (0.15 g/cm3) with organic texture and high porosities values (up to 86.2%) were found in samples with the highest organic material content in the floodplain soil. For this field experiment, flood plains soils (characterized as basin gley soils) presented high volumetric instability with a retratibility of 67.49% and higher water storage capacities compared to riparian stands soils (characterized as fluvic neosoils).

---

[*] Corresponding author: Valdemir Antonio Rodrigues. Departament of Forest, Faculty of Agricultural Science, University Estadual Paulista - FCA/UNESP - Botucatu - São Paulo - Brazil. E-mail: valdemirrodrigues@fca.unesp.br.

Keywords: soil, water, density, porosity, riparian jungle

Introduction

Vegetation consumes water according to their potential for stem development and the amount of available soil water surrounding the scope of its roots. Therefore, each plant species in its different stages of growth presents dynamics of soil water consumption by evapotranspiration, and the remaining water is stored according to soil texture and over time it will be available for the water table.

The presence of water in the soil is fundamental for the majority of its natural physical, biological and hydrological processes involved. This is very common in the specific case of the soils of riparian jungle (in Portuguese, mata ciliar) where the presence of water is linked to their natural processes. Riparian jungle ecosystems include vegetation that is settled in river banks. This vegetation plays an important role in the control of detrimental effects of the river rises and helps for the establishment of river banks. Riparian jungle soils are subject to frequent floods. In the soils of the riparian jungle the biological diversity and water storing is maintained according to (Penczak et al., 1994). In these soils, the saturated zone of the microbasin is extended during the rainfall season according to (Lima and Zakia 2000).

In riparian jungle, several types of soils are included. Fluvisoils is a type of soil that is very common in riparian jungle. Fluvisoils are included in all the continents and climates. With an extension of 350,000,000 ha, these soils supose more than a half of the surface of the tropics: Amazonian basin, the plain of Ganges-India, the Chad lake in Central Africa, Brasil, Paraguay and Argentina. Gley soils are another type of soil included in riparian jungle soils. Gley soils, with a surface of 720,000,000 ha, are azonal soils and are presented in the majority of the climates. These soils are extended from subartic areas in Siberia (Russia), Canada, Alaska, damp temperate soils and subtropical in China and Bangladesh and in tropical zones: Amazones, Africa and Asian Southeast, are reported for (IUSS-WRB 2007). The majority of soils are not expansives. Their volume is not increased when water is absorbed.

Moreover, some types of Argilic soils (with a high quantity of organic matter) are expansives. In these soils, the volume and the soil moisture is increased similarly according to (Prakash and Sridharan 2012). In these soils, the more soil moisture is decreasing the more contraction they suffer. This is a reversible phenomenon. The phenomenon of the volumetric instability of the soil (contraction and expansion) happens because of the change of humidity's level. This phenomenon is complex and influenced by several factors, such as the type of the soil, the climatic conditions and the tension states (Sartori et al., 1985; Tariq and Ferreira 1993).

There are several conditional factors linked to soil (water retention, air permeability, distribution of clay particles, porosity, mineralogic orientation, cimentation, estratigraphic profile, thick of soil and litological discontinuity, etc.) according to (Hunt 2003; Jalbert 2003) that were influencing its expansive potential. Soils that change in volume when they are flooded by water, as collapsible as expansion, demand an extra care regardless of the uses: agricultural, engineering work or both. This performance is inevitable because it is inherent to the soil according to (Ferreira 2009).

Retraction of a soil is a dependent variable of physical properties as density, porosity and water amount. A soil of a basin retains great amounts of water. A contraction is suffered by these soils when they begin to dry. When water is added soil begins to expand. (Rezende and Ferraz 1988). Another aspect that has to be observed is that, in expansive or retractive soils, density is altered because of volume variation. A dry volume of soil has to be taken in account in order to determine the density. The case of the volume of a humid soil is different. In most occasions, this aspect is not advertised. For this reason, wrong results for porosity and some doubts about evaluation of density could be generated.

High porosity (66%) evaluating physical soil properties are found in watershed friable soils. They developed some works about hidrogeology of the Paraiba do Sul river basin in a case study of mineralizing areas according to (Diniz 2010). The study of soil water movement is very complex because of the heterogeneity of the soil characteristics as density and size, the form and disposition of holes, texture, structure, specific surface, proportion of organic matter, iron oxides, hydraulic properties and other factors according to (Jorge 1985; Ferre 2002). The soils that present expansion and contraction an adequate volume for density calculations has to be considered. In this case the real volume of soil has to be considered. Additionally, it has to be taken into account with its volumetric contraction.

Vegetation is fundamental in the maintenance of the conditions of porosity of soils. This factor and the added organic matter generation are basic conditions for root development in the soil. Previous conditions of the use of soil could, in general, modify the characteristics of infiltration according to (Lima 1986).

Basin soils present a high quantity of organic matter and they suffer expansion and contraction. Organic soils present low values of density and high porosity. In this case, these soils change to store water and it contributes to the hydrological process of the watershed.

In the micro basins, water sustainability has to be a goal of preservation that is needed for the maintenance of ecosystems according to (Rodrigues 2011; Rodrigues 2014). An adequate soil and water management could decrease the degraded zones. The physic characteristics as expansion and contraction of soils control the water percolation in soil. A water storing environment is formed during rainfall season. The contraction of soil and the water outflow are developed along the dry season.

This work aims to study the relationship among soil moisture, density, porosity and soil contraction of basin soils, riparian stands and its capacity for water storing. The work is organized as follows: Section 2, the methodology is described in detail, Section 3 offers some results, and finally, Section 4 is devoted to the conclusions.

Material and Methods

Location of Study Area

The study area is located in the experimental farm of the Faculdade de Ciências Agronômicas-UNESP, São Manuel, São Paulo, Brasil. The hydrographical microbasin is cincunscribed between the next geographical coordinates: 22°46′27″ to 22°47′09″ of south latitude and 48°33′24″ to 48°35′34″ of west longitude. The average altitude is 770 m.

Soils and Climate of the Experimental Area

Two types of soils are next to one of others. Soils have a sandy texture. They present different characteristics: gley soil is over the basin, and fluvic neosoil is over the riparian stands. Basin presents a gley soil with 156 g/dm^3 of organic matter, and second (IUSS-WRB 2007) gley soil has a wide strip of non-consolidated materials, mainly fluvial, with basic to acid mineralogy, in a depressed area, low position in the land and with surface phreatic water. In the riparian stands, a fluvic neosoil with 48 g/dm^3 and organic matter is located. This soil is developed in genetically young alluvial deposits. The climate of the São Manuel, São Paulo, according to (Cunha and Martins 2009) was classified as Cwa, template hot climate (mesotermic) with rainfalls during summer and dry in winter. The average temperature of the hotest month is over 22°C.

Collecting Data of Soil Samples

Collecting data were extracted with volumetric rings of 100 cm^3, in vertical direction in two trenchs at 0, 30, 60 and 100 cm deeper, with three repetitions per layer. 24 samples were obtained along two profiles of soil of the riparian stands and the taboa basin (Typha domingensis). In these unaltered samples, the physical characterisitcs as soil moisture, aparent density, density of particles, volumetric shrinkage and water storing were obtained. The samples were collected 60 cm deeper of the soil profile.

Soil Moisure Determination

Soil moisture refered to dry mass was determined considering humid soil mass (mu). This parameter is obtained by weighing of samples and refered to dry soil mass (ms). Dry soil mass is obtained by drying in heater at 105°C constant mass included. Soil moisture refered to dry mass was obtained by the Equation 1. where: Ubs = [(mu-ms)/ms] × 100. Soil moisture referred to the volume was obtained multiplying soil density by soil moisture referred to dry mass. Equation 2. where: $\theta = \rho \times$ Ubs. Analysis was obtained according to (Embrapa 1999).

Determining Density, Density of Particles and Soil Porosity

Soil density (g/cm^3) was calculated by the ratio between dry soil mass (g) and the total volume of humid soil (cm^3). Moreover, for basin soils, a correction in the determination of density is applied, because of the volumetric shrinkage of this soil. In this case, the volume of dry soil is taken into account. Density of soil particles was obtained by the relation between dry soil mass (g) and the volume of the mineral and organic particles (picnometer method).

Total porosity of soil (Ps) was calculated using the average values of soil density (ρ_S) and the density of soil particles (ρ_P). by Equation 3. where: Ps = $[1 - (\rho_S/\rho_P)] \times 100$. In (g/cm^3). The according to (Klar 1988). Volumetric soil moisture (θ) is the rate between water volume and soil volume. This parameter is determined by the product of soil moisture and soil density according to (Klar 1988).

Volumetric Contraction Determining

Soil samples after retraction maintained an approximately cilindric format with some imperfections. These samples were considered as a cylinder in order to obtain the volume. According to (IUSS-WRB 2007), lineal expansion and contraction coefficient is a measure of soil expansion and contraction potential. This is defined as the relation between the size of the sample in saturation moisture and the size of the dry sample.

In this study, the contraction in the base of saturated and dry volume of the sample was determined. The high of the sample was determined by four determinings in different points and the average was calculated. The ratio of the sample was obtained from circle perimeter of the sample obtained with a line around a circumference of the sample in three different positions. Once the sample perimeter was obtained the average was determined and calculated the ratio of the sample dividing the perimeter by 2π.

According to the relation $Vs = (Pi)\ \pi R^2 h$, the dry volume of the samples was determined. Contraction or shrinkage of soil was determined using Equation 4, according to (Rezende 1997; Rezende 1998). $Rv = [(Vsat - Vs)/\ Vsat] \times 100$. Where, Rv = maximal volumetric contraction (%); Vsat = volume of satured soil (cm^3); Vs = dry soil volume (cm^3).

Quantification of water storage along the soil profile was determined 60 cm deeper.

The soil moisture was based on volume by the high of profile of soil every 10 cm, according to Equation 5, according to (Reichardt 1985). $\Delta s = (\theta \text{ x } h) \times 10$. (mm). Where, Δs = soil water storing (mm); θ = soil moisture referred to volume (cm^3/cm^3); h = soil profile height (cm).

Statistical Analysis

Results were analyzed using the software SAEG, version 9.1, UFV, 2007. For the variance analysis of the soil water storing the Tukey test was used, with an average comparison of 5% of significance. The porosity results related to average was expressed by the standard deviation and variance.

Results and Discussion

Contraction of Basin Gley Soils

Expansion and contraction of a soil are linked to other physic properties as, for example, the density and absorbed quantity of water during the drying (Figure 1a). In non-expansive soils, the volume is not increased when the water is absorbed and the contraction is not suffered when they loose soil moisture (Figure 1b).

In the basin soil, the volume of a dry soil sample is the volume of the volumetric ring. The reason of this is that basin soil suffers contraction after losing soil moisture like drying and after, with water addition, and sample returns to expand (Figure 1a). In the riparian stand soil the volume of a dry soil sample is equal to the volume of a volumetric ring. This soil suffers contraction and samples will remain with the same volume (Figure 1b).

The parameters to obtain the soil contraction were: R = average ratio of dry samples (cm); H = average height (cm); Vs = average volume of the dry sample (cm^3); m = dry soil mass (g); ρ as = obtained density by the ratio between dry mass and ring volume; ρac = obtained density obtained by the ratio between dry mass and the volume of dry soil (g/cm^3); ρp = density of particles (g/cm^3); Rv = volumetric contraction (%) in Table 1.

Basin soil presented a huge contraction with an average value of 67.5%, with a standard deviation of 4.41 and a variation coefficient of 20.04% (Table 1).

a b

Figure 1. Figure 1A presents the samples of basin gleysoil after drying in a heater. The volumetric contraction is evident. Figure 1B presents the samples of fluvic neosoil of riparian stands after drying in a heater. In this soil there was not contraction.

Table 1. Contraction of basin soil of Paraíso river Fazenda São Manuel

Sample	R	H	V_s	V_c	m_s	r_{as}	r_{ac}	r_p	R_v	R_v	$\bar{x}$ DP*	CV**
	(cm)	(cm)	(cm^3)	(cm^3)	(g)	(g/cm^3)	(g/cm^3)	(g/cm^3)		(%)		(%)
1	2,1	2,4	34,5	101,9	12,5	0,12	0,36	1,22	0,66	66,2	4,27	21,9
2	2,2	2,4	37,8	101,9	15,0	0,15	0,40	1,22	0,63	62,9	4,64	21,5
3	1,9	2,4	27,5	101,9	17,6	0,17	0,64	1,22	0,73	73,0	4,46	19,9
4	2,0	2,4	29,5	101,9	15,7	0,15	0,53	1,22	0,71	71,0	4,33	18,8
5	2,0	2,3	30,2	101,9	14,1	0,14	0,47	1,22	0,70	70,4	4,51	20,4
6	2,2	2,7	39,2	101,9	14,9	0,15	0,38	1,22	0,62	61,5	4,22	17,8
Average	2,1	2,4	33,1	101,9	15,0	0,15	0,46	1,22	0,67	67,5	4,41	20,0

*Standard deviation, **Variation coefficient.

Density of basin soil was 0.15 g/cm^3 with organic texture, the density of particles was 1.22 g/cm^3 and a high value of porosity was 86.2% (Table 2). It is verified that the basin soil is retractive, organic and presents a high porosity when it is compared with a riparian stand.

Results of physical-water soil properties of a basin soil of a riparian stand are: moisture mass (mu), dry mass (ms), moisture refered to dry mass, percentage and presents the samples of fluvic neosoil of riparian stands after drying in a heater. In this soil there was not contraction. Table 1 Contraction of basin soil of Paraíso River Farm São Manuel.

Figure 1a presents the samples of basin gley soil after drying in a heater. The volumetric contraction is evident; Figure 1b not percentage (Ubs), soil density (ρ), soil moisture refered to volume (θ), soil water storing (ΔS, mm) and soil porosity (P), are referred in Tables 2 and 3.

The surface of basin gley soil presents high values of porosity, and the average value is 87.9%, with an average standard deviation of 0.32 and the variation coefficient is 0.10%. The variance analysis was significant 5% for the water storing in a limited layer of soil that does not present significant differences with Tukey test at 5% (Table 2), when it is compared with the average data of the riparian stand of 29.8% of porosity. This difference could be caused to a major quantity of organic matter in the basin. The total amount of stored water in the basin for average dates was estimated at 60 cm deeper. This value was 588.7 mm. The fluvic neosoil of the riparian stand presents a minor porosity when it is compared with the basin gley soil. Average data of holes in the surface is greater when it is compared with other samples in deep. This is possible to be caused to a greater quantity of organic matter from vegetation in this superficial layer.

Soil density in the surface samples over the riparian stands was 1.30 g/cm^3 and, in profile average value was 1.48 g/cm^3 with heavy texture. In these samples the density of particles of soil was 2.10 g/cm^3. The surface porosity was 38.1% and in the soil profile was 29.5%, with an average standard deviation of 2.4 and a variation coefficient of 7.73%. The variance analysis was significant 5% for water storing in the limited layer presented and averages are not statistically different with the Tukey test at 5% (Table 3).

Table 2. Hydrophysical analysis of basin soil of Paraíso River Farm São Manuel

Profile	m_u	m_s	U_{bs}	U_{bs}	ρ	θ	P	Δ_s	DP*	CV**
Soil (cm)	(g)	(g)	(%)	-	(g/cm³)	(cm³)	(%)	(mm)	$\bar{x}$	(%)
0	111,6	15,3	626,8	6,27	0,15	0,945	86,3	37,8 c	0,3	0,09
30	115,1	15	666,8	6,67	0,15	0,982	86,6	39,3 b	0,34	0,12
60	115,7	15,3	654,7	6,55	0,15	0,985	86,3	39,4 ab	0,3	0,09
100	118	15,9	640,1	6,4	0,16	0,998	85,8	40,1 a	0,34	0,12
Average	**115,1**	**15,4**	**647**	**6,47**	**0,15**	**0,98**	**86,2**	**39,2**	**0,32**	**0,1**

Test of Tukey at 5%, *Standard Deviation and **Variation Coefficient.

Table 3. Hydrophysical analysis of riparian stand soil of Paraíso river Farm São Manuel

Profile	m_u	m_s	U_{bs}	U_{bs}	ρ	θ	P	Δ_s	DP*	CV**
Soil (cm)	(g)	(g)	(%)	-	(g/cm³)	(cm³)	(%)	(mm)	$\bar{x}$	(%)
0	172	132,5	29,8	0,298	1,3	0,387	38,1	15,5 ab	4,82	23,2
30	194,5	157,5	23,4	0,234	1,54	0,362	26,7	14,5 b	1,41	1,98
60	197,5	156,6	26,2	0,262	1,53	0,401	27,1	16,0 ab	1,51	2,29
100	201	155,7	29,1	0,291	1,53	0,443	27,1	17,7 a	1,85	3,43
Average	**191,2**	**150,6**	**27,1**	**0,271**	**1,48**	**0,4**	**29,8**	**15,9**	**2,4**	**7,73**

Test of Tukey at 5%, *Standard Deviation and **Variation coefficient.

Soil surface of riparian stand presents a major porosity when it is compared with the average data of profile with 100 cm deeper. This could be caused to organic matter from the plant residuals. The total amount of stored water in the soil of riparian stands in average data was estimated at 60 cm deeper. The result was 240.8 mm.

The difference between these two types of soil (basin gley soil and fluvic neosoil) of the riparian stand could be observed with a minor density and a major quantity of organic matter. According to the water storing in the soil of two environments the estimation of the water table was determined at 60 cm deeper. In the basin soil was estimated in (588.7 mm). In the riparian stand was estimated in 240.8 mm.

In this way, the basin presents an increment of 144.5% when it is compared with the riparian stand soil. The storing capacity of the basin soil is demonstrated. For this reason it is clearly that basin is a remarkable environment as natural water storing. This water could be available for the rivers in critical seasons, as long as dry seasons.

This aspect results in the importance of vegetation Typha domingensis in order to improve the water retention capacity of basin gley soil. It is possible to a firm that basin soil, as gley soil type has an excellent quality about the physical and chemical aspects. These properties are related to the great ratio of organic matter. This organic matter is provided by its vegetation and the accumulation of sediments of the edge in the riparian stand vegetation. It is notable the mutual help of these vegetations to control the temperature, erosion and the improvement of fertility of two soils. Additionally, the edge effect of the riparian stands contributes to the adding of the organic matter, sediments and non-consolidated materials for the basin gley soil.

The environment of the basin contributes to the water availability for river during the dry seasons. This is an suitable microclimate for the environment and vegetation of riparian stands.

CONCLUSION

Basin gley soil presents an extremely high value for the volumetric contraction and a high natural volumetric instability. This is a physical property that implies a great capacity of water storing.

Fluvic neosoils of riparian stands do not present volumetric contraction. Although these two types of soils are located nearby, they present very different physical properties.

Low density values and high porosity values of the basin gley soil imply a high content of organic matter in the soil. This occurs in fluvic neosoil of the riparian stands.

The basin gley soil presents a higher value of stored water quantity than the riparian stand soil, because of the high values of the organic matter and the high volumetric instability.

REFERENCES

Cunha, A. R.; Martins, D. 2009. *Classificação climática para os municípios de Botucatu e São Manuel*, SP. Irriga, Botucatu, v. 14, n. 1, p. 1-11.

Diniz, H. N.; Galina, M. H.; Batista, G. T.; Targa, M. S. 2010. Hidrogeologia da várzea do rio Paraíba do Sul: estudo de caso de uma área de mineração no município de Tremenbé, São Paulo, Brasil. *Ambi-Água*. Taubaté, v. 5, n. 3, p. 76-107.

EMBRAPA. 1999. Sistema brasileiro de classificação de solos. Brasília, DF. p. 412.

Ferre, T. P. A.; Nissen, H. H,; Simunek, J. 2002. The effect of the spatial sensitivity of TDR on inferring soil hydraulic properties from water content measurements made during the advance of a wetting front, *Vadose Zone J.* 1(2) p. 281-288.

Ferreira, S. R. M.; Ferreira, M. G. V. X. 2009. Mudança de volume devido à variação do teor de água em um vertissolo no semiárido de Pernambuco. *Revista Brasileira de Ciência do Solo*, Viçosa, MG, n. 33, p. 779-791.

Hunt, A. G.; Gee, G. W. 2003. Wet-end deviations from scaling of the water retention characteristics of fractal porous media, *Vadose Zone J.* 2(4) p. 759-765.

IUSS GRUPO DE TRABAJO WRB. 2007. Base referencial mundial del recurso suelo: un marco conceptual para clasificación, correlación y comunicación internacional. Roma: FAO, (Informes sobre Recursos Mundiales de Suelos, 103).

Jalbert, M.; Dane, J. H. 2003. A handheld device for intrusive and nonintrusive field measurements of air permeability, *Vadose Zone J.* 2(4) p.611-617.

Jorge, J. A. 1985. *Física e manejo dos solos tropicais*. São Paulo: Instituto Campineiro de Ensino Agrícola. p. 328.

Klar, A. E. 1988. *A água no sistema solo-planta-atmosfera*. 2. ed. São Paulo: Nobel. p. 408.

Lima, W. P. 1986. *Princípios de hidrologia florestal para o manejo de bacias hidrográficas*. Piracicaba: USP/ESALQ. p. 241.

Lima, W. P.; Zakia, M. J. B. 2000. Hidrologia de matas ciliares. In: Rodrigues, R. R.; Leitão Filho, H. F. *Matas ciliares: conservação e recuperação*. São Paulo: EDUSP/FAPESP. 3, p. 33-44.

Penczak, T. Agostinho, A. A. Okada, E. K. 1994. Fish diversity and community structure in 2 small tributaries of the Parana River, Parana State, Brazil. *Hydrobiologia*, 294(3) p. 243-251.

Prakash, K.; Sridharan, A. 2012. Sediment volume tests for expansive soil identification and classification. *Journal of Testing and Evaluation, Conshohocken*, v. 38, n. 6, p. 1-7.

Reichardt, K. 1985. *Processos de transferência no sistema solo-planta-atmosfera*. 4. ed. Campinas: Fundação Cargill. p. 466.

Rezende, M. A. 1997. Uma abordagem não convencional sobre as principais características físicas da madeira, com ênfase para retratibilidade, massa específica e técnica de atenuação da radiação gama. 1997. 138 f. Tese (Livre - Docência) - Instituto de Biociências, Universidade Estadual Paulista, Botucatu, São Paulo, Brasil.

Rezende, M. A. 2003. Wood shrinkage of *Pinus caribaea var. hondurensis* and *Eucalyptus grandis* and their relationships with moisture content and specific gravity. *Scientia Forestalis*, Piracicaba, n. 64, p. 120-127.

Rezende, M. A.; Ferraz, E. S. B. 1988. Retratibilidade volumétrica e densidade aparente da madeira em função da umidade. *IPEF*, Piracicaba, n. 39, p. 33-40.

Rodrigues, V. A. 2014. Avaliação dos processos hidrológicos em microbacias hidrográficas.125f. Tese (Livre - Docência) - Universidade Estadual Paulista. Faculdade de Ciências Agronômicas. FCA – UNESP. Botucatu, São Paulo, Brasil.

Rodrigues, V. A. 2011. Análise dos processos hidrológicos em modelo didático de microbacias. *Revista Cient. Eletrônica de Engenharia Florestal*, Garça, v. 17, n. 1, p. 1-15.

In: Soil Management
Editor: Manuel Esteban Lucas Borja

ISBN: 978-1-63483-274-8

Chapter 10

SUSTAINABILITY OF SOIL, WATER AND FOREST IN TROPICAL FOREST

Valdemir Antonio Rodrigues[1,*], Manuel Esteban Lucas-Borja[2], José Maria Tarjuelo[2] and Antonio Ruiz Canales[3]

[1]Department of Forest, Faculty of Agricultural Science, University Estadual Paulista - FCA/UNESP – Botucatu, São Paulo, Brazil

[2]Department of Agroforestry Technology and Science and Genetics, School of Advanced Agricultural Engineering,
Castilla La Mancha University, Albacete, Spain

[3]Escuela Politécnica Superior de Orihuela, Universidad Miguel Hernández de Elche, Orihuele, Spain

ABSTRACT

The sustainability of the system soil-forest-water involves essential variables such as soil conservation and its productive potential, the maintenance of plant's and animal's biodiversity, the maintenance of the hydrological processes, the dynamic of surface and groundwater, the intensity of human activities and the use and management plans related to the forest. These are fundamental in the biodiversity and water resources conservation. Considering that soil conservation is the basis of production, forest is essential in soil protection. The sustainability of water in watersheds is provided by both elements. The sustainable development is consolidated by this interactivity. This depends on the boosting of environmental values about nature conservation, management and use of natural resources in a sustainable way for the preservation of life on the planet.

Keywords: sustainability, soil, water, forest

---

* Corresponding Author address, Email: valdemirrodrigues@fca.unesp.br.

INTRODUCTION

The sustainability of natural resources depends fundamentally on the intensity of human activities and the management and soil, water and forest. These are specifically: soil conservation and its productive potential of plant and animal foods, the maintenance of bio diversity of tropical forests, the rational management of forests in the production of wood, fruits, seeds and medicinal essences, the maintenance of hydrological processes, conservation of springs, the dynamics of water and the recharge of the aquifers.

The report of the United Nations Food and Agriculture (FAO, 2011) indicates that 25% of soils in the world are in an advanced stage of degradation, endangering production systems and the quality of water resources. Biodiversity is destroyed by forest fires and deforestation and this causes unrepairable degradation of the ecosystems.

Among the biomes, tropical forests have the greatest biodiversity. The Atlantic Forest (Brazilian Biosphere Reserve- UNESCO) is one of the main two shelters of biodiversity across the planet. However, this forest is included among the 17 most threatened ecosystems in the world (Soares et al. 2006). The large forest biomes of Brazil (Amazon rainforest, Cerrado vegetation, Caatinga, Atlantic forest, broadleaved semi-deciduous forests, Araucaria forest and Mato Grosso wetland) are plant and animal biodiversity reservoirs. Therefore, the areas susceptible to degradation should be preserved. The conservation and management of soils are essential conditions in agricultural productivity, livestock and forestry. The soil is the basis of production, combined with the techniques of nutrition, irrigation and the genetic improving of seeds and clones, where an increase in agricultural productivity, animal and forest is included.

The presence of essential elements as water, carbon, energy, soil, forest and others is required for the existence of the life on the Earth. Both urban and rural farming areas the numerous uses of water are indispensable in a wide range of activities. During the Eco-92 in the city of Rio de Janeiro (Brazil) the organization of the United Nations (UN) determined that sustainable development should be: economically viable, environmentally sound and socially just. Sustainability is a process of rational use of natural resources without degradation, preserving resources for present and future generations.

The decade of water from 2005 to 2015 was declared by the UN. The subjects of this decade were Water for Life, Water for Ecosystems, Water for People, Ecosystems for People. When it comes to water, some challenges show up like: the fact that water consumption in several sectors is not sustainable, what depletes groundwater reserves and degrades ecosystems; urban floods and scarcity of public supply of water; chemical contamination of river water; water pollution with urban and industrial effluents. Nearly 300 million people in Africa are now without access to clean water. One of the millennium development goals is to reduce this number to 150 million people by 2015.

The watershed of Mount Catskill provides clean water for the State of New York – US. Due to conservation of riparian ecosystem, soil microorganisms and the riverside vegetation, natural processes of water filtration are provided. The investment of $ 1-1.5 billion in the purchase and protection of mountain watersheds Catskill was chosen by the State of New York rather than the investment of $ 6-8 billion in an artificial plant filtration.

The “Serra dos Órgãos” National Park is a protected area with 20,024 hectares of biodiversity protected with dense rain forest in the Atlantic Forest, home to over 2,800 species of plants, 369 species of orchids, more than 100 of bromeliads, the juçara palm tree (Euterpe edulis), which the palm is extracted from, the jequitibá pink (Cariniana estrellensis), among other species in the mountainous region of Rio de Janeiro - Brazil. The park is home of 462 species of birds, 105 mammals, 103 amphibians and 83 reptiles, including 130 endangered animals and many endemic species according to the SOS Atlantic Forest (2015). Biosphere Reserve is a model adopted internationally in an integrated, participatory and sustainable management of natural resources (MMA, 2015). It is recognized by the program "Man and the Biosphere (MAB)" of the United Nations Educational, Scientific and Cultural Organization (UNESCO). In Brazil, in the biosphere reserve of the green belt of the city of São Paulo and gardens and forest reserves of the Instituto Florestal, the plans of forest use and management have as purpose the biodiversity and water resources conservation. Therefore they are examples of: Water for Ecosystems, Water for People, Ecosystems for People. This paper proposes to discuss the importance of soil sustainability system, forest and water. It is known that the subject is an on-going challenge of contemporary society and, in order to contribute, presents some aspects of sustainability.

Forest Soil Link

Soil is the support base and the forest ecosystem protection. The dynamic and sustainability of water in watersheds is promoted by both aspects. All terrestrial, aquatic ecosystems and flows in the oceans are re-supplied by water. The inadequate human actions on the ground allow the conduction of the organic matter, sedimentation and siltation of rivers. The organic matter is the result of perfect interaction between forest biomass, soil organisms and water, which must be preserved for nutrient cycling.

The conservation of the existing plant and animal diversity in nature, life in different stages of the life cycle is provided by natural forests. All of them are playing important roles in that biome, ensuring life in ecosystems, producing medicinal essences, food, energy, biomass, and other direct and indirect benefits for human disposal in a sustainable way. In the understory forest seeds are spread on land by means of wind, rain and animals, where there germination, growth and competition between plants. An example of this is the presence of social insects as the ants (*Atta* spp) living in the soil in society or community in full interactivity with vegetation. Another example is the presence of flowers in the fields and forests, bees (*Apis mellifera*), bats (*Chiroptera*) and kiss flowers (spp Hummingbird), among others, are pollinators that feed back into this dynamic with the essential environmental services in sustaining biodiversity. Environmental forest services such as water conservation, biodiversity, carbon sequestration, oxygen production and scenic beauty of the landscape are the resultant vast values of this ecosystem relationship: soil, water, animal and forest.

Water is consumed by vegetation according to their potential for stem development and amount of available soil water surrounding the scope of its roots. Evapotranspiration is the mechanism by means plants consumes ground water. Photosynthesis is closely linked to the process of sweating and synthesis of nutrients. This process occurs in the chloroplasts of the leaves of plants in the presence of light, where CO_2 is converted to O_2. Therefore, each plant species in its several stages of growth a dynamic consumption and water storage capacity are presented according to the soil texture and the microclimate.

Forests are the main sources of organic matter, as well as by the deposition of branches and leaves, and by the contribution made by the roots. In the soil, organic matter is found in different stages of decomposition. Content and decomposition of organic matter are the result of several factors such as the type of vegetation, topography, climate and time (age). In soil, accumulation or destruction are submitted to the effect of the activity of microorganisms, which are affected by moisture conditions, aeration, temperature and the environmental reaction (pH), and the supply of essential nutrients and energy sources (Kiehl, 1979). According to the author, organic matter is defined as the vegetal and animal material in raw state or in the process of decomposition or humidified and soil humus. All the vegetal and animal wastes can undergo a process of biological decomposition under the influence of microorganisms. In this way, humus is characteristically constituted around 50-80% of the total organic matter. Moreover, this substance includes physical, inherent chemical and physico-chemical properties, between dark brown and black in colour, high contents of lignin and protein, carbon-to-Nitrogen ratio (C/N) around 10: 1, however without a definite breakdown. The C/N gives information about the state of humidification of organic matter in the soil because humus has a C/N value ranging from 12: 1 to 8: 1 with an average of 10: 1 (Kiehl, 1979).

When the soil organic matter is transformed to humus together with the clay fraction is defined as "colloidal complex of soil". The decomposition of biomass in organic matter is a process of nutrient cycling. Throughout the soil profile, the roots of vegetation and soil horizons can be observed. The first organic layer is with microorganisms, and horizons: A, B, C, and R, where there is the rock matrix weathering and soil formation process (morphogenesis). Rain also contributes to the input process of nutrients in the soil.

The soil microbiota is constituted by a microscopic mixture of thousands and millions of bacteria, fungi, protozoa, etc. This could vary depending on the plant species, age and nutritional status of the plant, handling and environmental conditions, with soil enzymes and under certain physicochemical conditions. Soil microbiota plays an essential role in the processes of degradation, mineralization and oxidation-reduction of existing matter in the soil. This process makes it possible to maintain the soil fertility. For this reason a complete study of microbiological, biochemical and physicochemical properties will give an adequate idea of the functionality of soils.

WATER SUSTAINABILITY

Water is the source of life and one of the pillars of sustainable development. The Brundtland Report, "Our Common Future - Our Common Future" was widely discussed and known in the Eco-92, Agenda 21, UN conference in Rio de Janeiro (Brazil), 1992. In this document sustainable development is conceived as the development that satisfy the present needs without compromising the capacity of future generations to supply their own needs (UN, 1987).

The hydrologic cycle provides the water sustainability in the nature. Water is responsible for the maintenance of life on earth. The water cycle has nor beginning nor end. This is a closed loop. In nature, water is circulating in vapor state when is evaporated and rises into the atmosphere, forming atmospheric humidity. Water is in liquid state, when is condensed and falls through the precipitation (rainfall). Moreover, water is in solid form, present in glaciers, forming the ice caps of the globe. The major contribution of atmospheric humidity comes from the evaporation of water from rivers, lakes and oceans. Vegetation contributes to the atmospheric humidity by means the evaporation, although on a smaller scale.

Hydrological processes that provide sustainable water in watersheds (Figure 1) are: regional rainfall, the physiographic conditions of the watershed and the natural state of conservation of soil, human actions through the management and conservation of soil, water and forest. These processes are adopted in rural and urban areas of the inserted regions in the watersheds. The runoff is a hydrological process of interaction between rainfall and the physiographic conditions of watersheds and the flow is the result of this process. The water production in the source can be variable in time and space specifically in each region (Rodrigues, 2014). This depends on regular rainfall, the physiographic conditions and soil management adopted. The differences in the range of river flows between the two dry and wet periods depend on the forest cover and soil conservation management adopted by farmers. Human activities carried out in the management and soil conservation have direct influence on water infiltration of rainfall, the recharge of the aquifer and base flow. These provide greater water increase in the aquifer, and greater regulation of river flows in dry and rainy season.

The riparian ecosystem with the riparian forest (Figure 1) situated between cultivated areas and waterways acts as a "buffer zone or filter" is the interface between the aquatic and terrestrial ecosystems, second (Rodrigues and Carvalho, 2004).

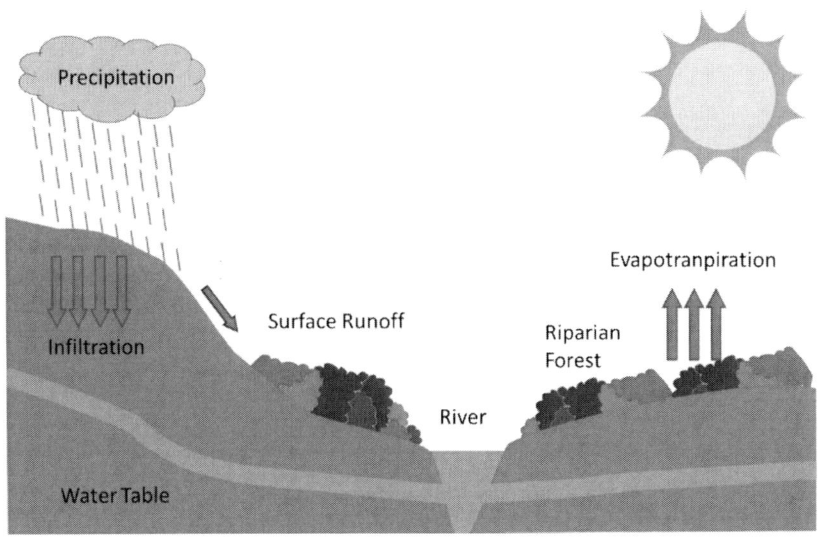

Figure 1. Hydrological processes in watershed.

In watersheds, in humid areas, where there is runoff, second (Rodrigues and Carvalho, 2004), is the transport of nutrients in solution, released the terrestrial ecosystem. Nutrients, crossing the riparian zone, are retained by absorption by the root system of the vegetation. Most of the suspended sediment in rivers comes from its own margins, confirming the importance of riparian vegetation as a stabilizing agent. Among other attributes (Rodrigues,

2014) points out the influence of the forest in the dynamics of water and recharging the aquifer.

According to Hamilton and Cassels (2003), the forest minimizes erosion and reduces the problem of sedimentation. They emphasize also that the sedimentation can make the water unsuitable for many uses: public, agricultural and industrial supplies. In this same line of thinking, sedimentation can reduce the navigation capacity of rivers, aquatic biodiversity and fisheries; reduce the capacity of reservoirs and increase friction in hydroelectric turbines.

Forest has an influence in the water dynamics and aquifer recharge. During the last decades, man has been irrationally exploring the forest with logging, ecosystem degradation and rivers pollution. The consequences are the reduction of the recharge of the aquifers. This fact is greatly changing the capacity in the production of organic matter and soil erosion, the accumulation of sediment into rivers. This is altering water quality, harming the sources of public water supply and other water uses as industry, agriculture and ecotourism. The water scarcity that affects millions of people in all regions of the world, the water contamination, the urban floods and the degradation of springs in the watersheds are main problems of today's society and are caused by the nature's destruction by humankind.

Protection of Forests

Around the 50% of the permanent preservation areas (PPA) of this watershed are protected and, according to Brazilian law, they are essential in the aquifer recharge and conservation of water resources. Therefore, the 50% of the PPA need to rebuild the forest, because of its given importance in regulating the hydrological system. In the riparian environment plant species are tolerant to wet, saturated or flooded soils, and perform important hydrological functions in the regulation of watercourses and the stability of the banks of the canals. This riparian environment provides a filtering system of the carried sediments by means of rainwater, reducing the pollution load in river water. This is essential for water quality and maintenance of aquatic ecosystems.

The remains of vegetation on laterals or slopes more than 45 degrees in their respective biomes have important hydrological and ecological functions with decreasing superficial flow and increased water infiltration into the soil. The vegetation of the Cuesta de Botucatu (Figure 2) and the Atlantic Forest biome, such as protected areas of the State Park of Serra do Mar, act as protectors of colluvial soil and act in the conversion of runoff into the ground. This helps in the areas of the recharge sources of watersheds and reduces sedimentation, erosion and siltation of rivers.

Therefore, the preservation of the front Cuesta is essential to the conservation of environmental services, especially water, soil and biodiversity, which enriches the landscape and feeds back the water in the waterfalls (Rodrigues, 2014).

Figure 2. Remnant vegetation of the watershed. Botucatu - São Paulo – Brazil.

Forest environmental services have a vital and strategic importance in the conservation of water for the future of humanity. For instance, vegetation in the areas of legal reserve is the 20% of the area of the farms in the State of São Paulo (Brazil). This native vegetation plays an important role as a germplasm bank, and a genetic and strategic reserve available to man. Moreover the riparian forests act in the protection of wildlife, biodiversity conservation, and water dynamics in the recharge of aquifers. It is noteworthy that the agroforestry zoning is an important tool for definition of protected areas and farming areas of production.

According to Johnson and Perrot-Maitre (2000), forest ecosystems provides to society four types of benefits related to water: quality water with low levels of nutrients and chemical pollutants; regulation of the surface and groundwater flow in different ways; to prevent or mitigate floods and landslides. The water supply in the regularization of the annual flow can, however, in some situations, increase the flow in the dry season, which benefits farmer's irrigation, water supply companies, hydropower and large users who rely on regular water flows in the dry season. The aquatic productivity is also affected by forest conservation. The conditions related to fishing, when it comes to quality and quantity, usually are associated with the characteristics "upstream" of the basins. As beneficiaries of this service are the practitioners of sport and commercial fishing, official agencies fishing and tourism industry.

The Amazon Forest has the greatest biodiversity on earth. Actually, there are still lots of unknown species. The flora counts with a variety of 30.000 species, about 10% of plants on the planet. There are about 5.000 tree species (bigger than 15cm in diameter), while in North America there are about 650 species of trees. The diversity of trees varies between 40 and 300 different species per hectare, while in North America ranges from 4 to 25. A WWF's survey, using UN data, shows that the average deforestation in the Brazilian Amazon is the world's largest. It's 30% more intense than deforestation in Indonesia, which is the second in the ranking of environmental devastation. The Amazon deforestation has grown, what affects soil, regional hydrology, the global carbon cycle, evaporation rates, loss of biodiversity, probability of bushfire and a possible reduction in regional rainfall. Official data from INPE show a high deforestation in the Amazon region. It has been eliminated approximately 570.000 kilometers of forest in the region, which corresponds to an area the size of France. The great challenge is to seek knowledge about the characteristics ecosystems of the Amazon and make suggestions about the sustainable development (Secco, et al., 2010). It is therefore concluded that soil conservation is the basis of production. The forest is to protect the soil and both provide the sustainability of water in watersheds. To promote environmental education to children, raising awareness about the importance of preserving nature and the appropriate management of natural resources, can contribute to keep the sustainable interaction between water, soil and forest for the benefit of society.

Conclusion

Biodiversity conservation of natural forests and the reforestation with native fruit species are fundamental in the conservation of water, soil, forest and sustainability of fauna with greater supply and environmental services for the availability of the humanity. Empowering and multiplication of environmental values to the community about the importance of nature conservation and use of natural resources sustainably are essential in preserving life on the planet. To preserve the soil around permanent preservation areas (PPAs) protects the springs of the rivers, as well as the preservation practices held in rural and urban environments are related to the reduction of erosion and water runoff. The impact of floods in the cities and in the riparian ecosystem is reduced. The result is consistency in the production of water from watersheds. The erosion, flooding and silting of rivers break the sequencing of the water in the hydrological cycle and destabilize the sustainability of water in watersheds that damaged the agricultural, livestock and forestry production. Containment systems of steep slopes such as cleaning, dykes in ravines, water impoundment boxes in higher yield points, the forest use and management plans are fundamental in the conservation of biodiversity, soil and water resources.

References

Hamilton, L.; Cassels, D. Hidrology overview. In: Dudley, N.; Stolton, S. (Ed.). Running pure: the importance of forest protected areas to drinking water. Washington: World Bank; WWF Alliance for Forest Conservation and Sustainable Use, 2003. p. 58-62.

Johnson, N.; Perrot-Maitre, D. Market-Based Instruments and Watershed Management: Overview. 2000. In: Workshop developing markets for environmental services of forests, 2000.Vancouver. Proceedings… Vancouver: British Columbia, 2000.

Kiehl, E. J. Manual de edafologia: relações solo planta. São Paulo: Agronômica Ceres, 1979. p. 262.

MMA. Ministério do Meio Ambiente. Reserva da Biosfera. 2015. Disponível em: <http://www.mma.gov.br/areas-protegidas/acoes-e-iniciativas/gestao-territorial-para-a-conservacao/reserva-da-biosfera>. Acesso em: 20 mai. 2015.

Organização das Nações Unidas para Agricultura E Alimentação. FAO. Escassez e degradação dos solos e da água ameaçam segurança alimentar. Brasília, DF, 2009.

Rodrigues, V. A. Avaliação dos processos hidrológicos em microbacias hidrográficas. Tese de Livre Docência. Universidade Estadual Paulista - Faculdade de Ciências Agronômicas. UNESP. Botucatu. São Paulo. Brasil. 2014. p. 125.

Rodrigues, V. A.; Carvalho, W. A. Análise morfométrica da microbacia hidrográfica do Córrego Água Limpa. In: Workshop em manejo de bacias hidrográficas, 8., 2004, Botucatu. Anais... Botucatu: Universidade Estadual Paulista, 2004. p. 144-163.

Secco, M. F. F. V.; Santos, J. B.; Fonseca, M. J. C. F. Biodiversidade da Amazônia. Prêmio Márcio Ayres. 2010. Disponível em: <http:// marte.museu-goeldi.br/marcioayres/index.php?option=com_content&view =article&id=10&Itemid=11 Acesso em: 20 mai. 2015.

Soares, P. V.; Agueda, W. L.; Martinelli, M.; Ananias, F. C. W. Reflorestamento conservacionista em bioma de mata atlântica, na bacia do ribeirão Guaratinguetá. In: RODRIGUES, V. A.; BUCCI, L. A. Manejo de microbacias hidrográficas: experiências nacionais e internacionais. FEPAF, Unesp, Botucatu, 2006, p. 80 – 87.

SOS Mata Atlântica. Série Parques do Brasil – Parque Nacional da Serra dos Órgãos.2015. Disponível em: <http://www.sosma.org.br/blog/serie-parques-brasil-parque-nacional-da-serra-dos-orgaos/>. Acesso em: 20 mai. 2015.

In: Soil Management
Editor: Manuel Esteban Lucas Borja

ISBN: 978-1-63483-274-8

Chapter 11

MANAGING SOIL EROSION IN SUB-SAHARAN AFRICA USING MODERN TECHNOLOGY

Lister Dube, Georgina Dede Arthur* and Kuben Naidoo
Department of Nature Conservation,
Mangosuthu University of Technology, Umlazi, Durban,
KwaZulu-Natal, South Africa

ABSTRACT

Human activities such as improper farming, logging, construction, overgrazing by livestock, off-road vehicle use, mining and deliberate burning of vegetation all accelerate the rate of soil erosion. The deposited sediments pollute water, air and affect the environment in many ways. Moreover, the increasing rate of human population induced changes is currently the major environmental concern causing soil degradation in comparison to natural factors. Techniques to conserve and stop persistent problems of land degradation by erosion are being developed, discovered adopted and implemented worldwide. The concept of managing the soil as natural resources by the people for their own benefit is critical in order to realize the full potential of its benefits. The involvement of communities in areas affected by erosion must be encouraged so that they are capacitated with skills to rehabilitate degraded areas and conserve soil. The aim of this chapter is to describe the extent of soil erosion and to discuss modern technologies used to mitigate these challenges.

Keywords: Sub-Saharan Africa, soil degradation, soil erosion, soil fertility, soil management practices, soil conservation

* Corresponding author: E-mail: Georgina@mut.ac.za. Tel: +27 31 907 7622, +27 724 732 437, Fax: + 27 862 737 797.

1. INTRODUCTION

According to the Independent Commission on Human Security the three main sources of environmental threats are consumption of fossil fuels and increased pollution; land degradation due to overuse, and the build-up of greenhouse gases (Hendrix and Glaser, 2007). Africa is endowed with a vast range of ecological environments. Most sub Saharan African countries are deep rooted in agriculture and ploughed land is left bare for several months after harvest until the next cultivating season. Erosion is defined as a "process which transfer soil from one location to another by wind, water, or other natural action and often accelerated by humans" (Pater and McReynolds, 2005). Overgrazing in some areas by livestock also destroys large expanses of grass thus denuding the ground and exposing it to agents of erosion (FAO, 2001). Most sub Saharan countries are in a transitional zone of becoming developed countries despite the fact that the African population continues to grow at higher rates than in other continents. Soil fertility depletion in sub Saharan Africa is still considered a major biophysical factor limiting per capita food production on the Continent (FAO, 2001; De Graaff et al., 2011, Sommer et al., 2013). In South Africa, over 70% of the lands are affected by varying degree of soil erosion (Pretorius 1998; Garland et al., 2000; Le Roux et al., 2008). Natural erosion occurs gradually while human induced soil erosion occurs at a faster rate (Pater and McReynolds, 2005). Erosion results in decline of soil fertility which refers to deterioration in soil physical, chemical and biological properties (Paustian et al., 1997; Giller et al., 1998; Tilman, 1999; Lal, 2009). Land degradation caused by soil erosion involves the loss of fertile topsoil and reduction of soil productivity, sedimentation of reservoirs increases suspended sediment concentrations in streams, with consequent effects on ecosystem health (Flügel et al., 2003; Lal, 2003). Other consequences include compaction which results in poor drainage, poor aeration and hard plain surfaces which cause runoff. In addition, soil erosion which includes sheet, rill, gully and land slide has severely compromised sustainable agricultural practices (Drechsel et al., 2001; Dotterweich, 2013).

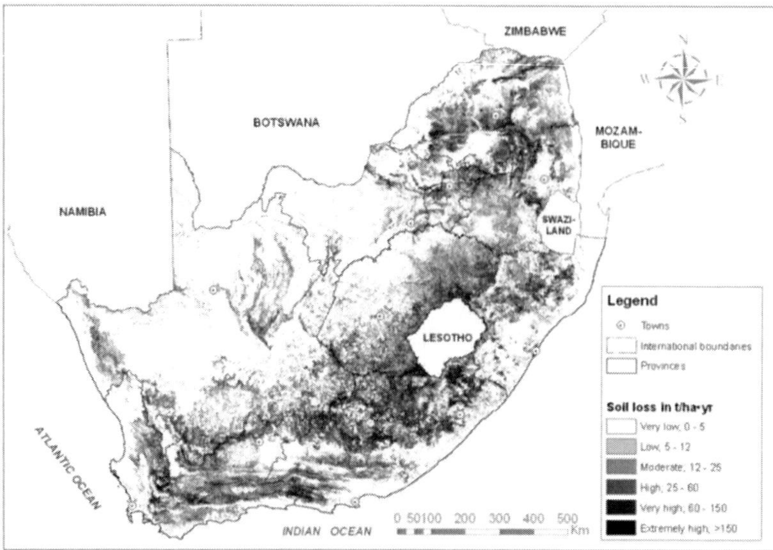

Figure 1a. Graphical representation of erosion in South Africa (Source Le Roux et al., 2008).

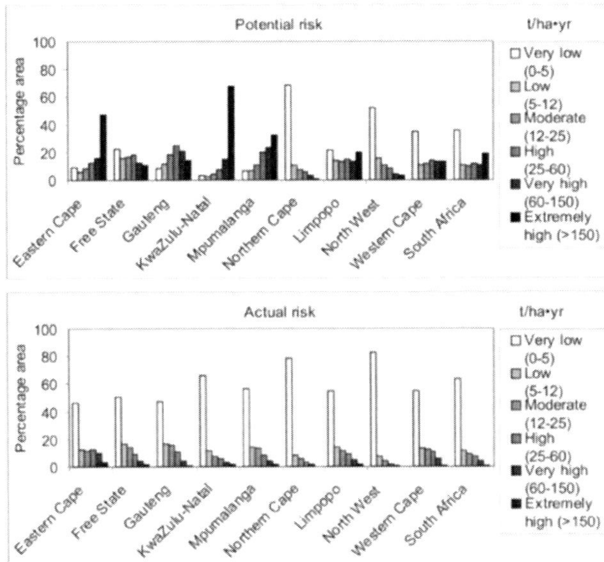

Figure 1b. Potential and actual erosion risk of each province expressed as a percentage (Source Le Roux, et al., 2008).

Human activities such as improper farming, logging, construction, overgrazing by livestock, off-road vehicle use, mining and deliberate burning of vegetation all accelerate the rate of soil erosion (Regan, 2011). The deposited sediments pollute water, air and affect the environment in many ways. The degree of erosion in South Africa is shown in Figures 1a and 1b. Most of the erosion occurs in agricultural farmlands away from the city centers and in rural areas where indiscriminate life-stock farming is a problem. The aim of this chapter is to describe extent of soil erosion and to discuss the modern technologies used to mitigate rehabilitation efforts.

2. Causes of Soil Degradation

Soil is a renewable resource which is composed of mineral particles formed over thousands of years from the breakdown of rocks. Soil is formed at a rate of only 1 cm every 100 to 400 years and it takes 3 000 to 12 000 years to build enough soil to form productive land (Department of Agriculture, 1999). To cope with the escalating demand for food it is essential to preserve the quality of the soils and ensure their sustainable use. Soil erosion is an on-going process worldwide and has produced river valleys and shaped hills and mountains. The causes of soil erosion are shown in Figures 2a and b.

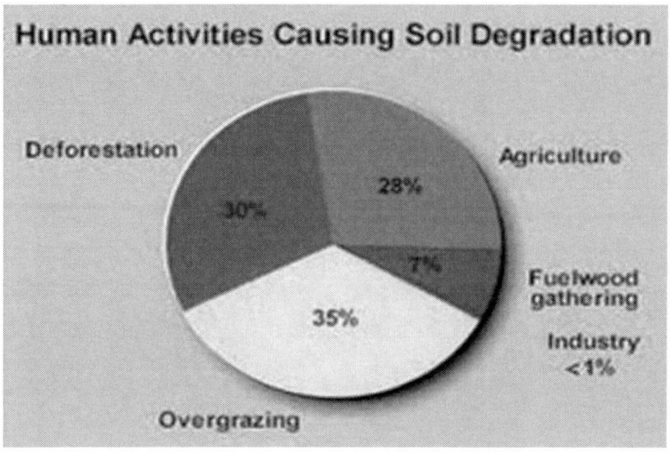

Figure 2a. Human induced changes to soil in rural areas (Source: http://haiti-sustainability.yolasite.com/water-sources.php).

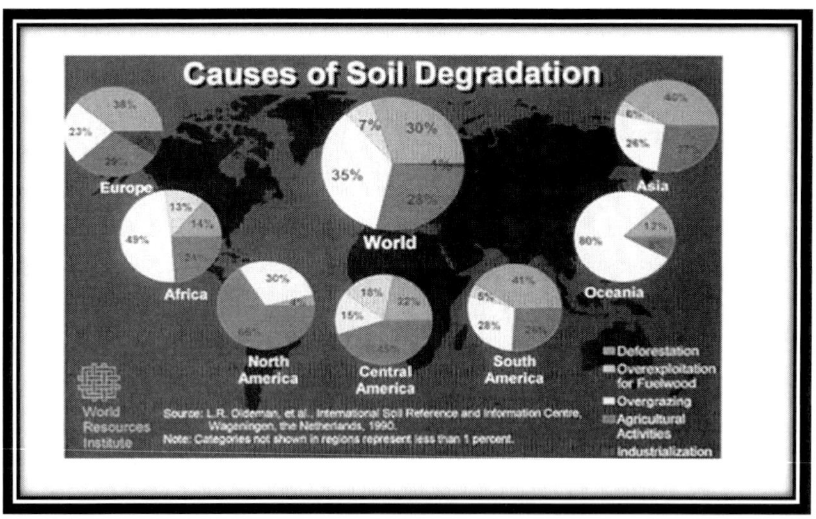

Figure 2b. The major causes of soil erosion worldwide (Source:http://www.globalchange.umich.edu/globalchange2/current/lectures/land_deg/land_deg.html).

3. Types of Soil Erosion

3.1. Sheet erosion is where a fairly uniform layer of soil is gradually removed over an entire surface area (Camp and Duagherty, 1997). This is caused by splash from raindrops with the loosened soil generally transported in rills and gullies and may go unnoticed until most of the productive topsoil has been lost. Figure 3a-3c shows the different types of erosion.

Figure 3. a) Sheet erosion, b) Rill erosion c) Gully erosion (Source: http:// awwatersheds.org).

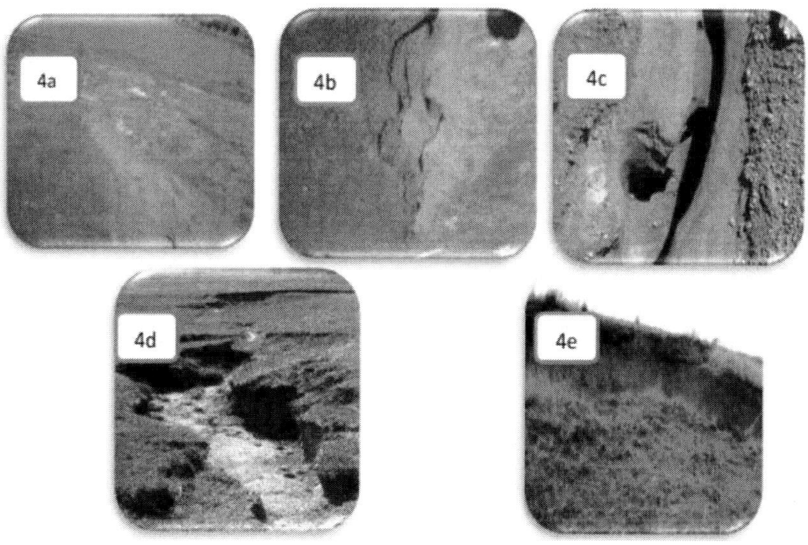

Figure 4a-e. Early stage of gully a, Gullies requiring immediate minimum action b and c, Serious and severe gully erosion requiring many resources for correcting d and e (Source: www://awwatersheds.org).

3.2. Rill erosion-occurs when soil is removed by concentrated water running through little streamlets, or headcuts. Generally, rill erosion is understood as the effect of flowing water exceeding a certain threshold of soil resistance (Knapen et al., 2007).

3.3. Gully erosion (Dongas) occurs when continuous serious rill erosion takes place by removing the soft soil, rocks, and expanding, as well as deepening and flowing together to make larger streams (Figures 4a-e show the development of gullies up to a level of looking like a stream. A rule of thumb is that if you can only fit your hand in it, it is a rill. If you can step in it, it is a gully.

3.4. Bank erosion is caused by water cutting into the banks of streams and rivers. It can be very serious in times of large floods and cause major destruction to property. The soil instability of fully banks, usually associated with seepage of ground water, leads to sloughing and slumping (caving-in) of bank slopes (Ritter, 2012)

4. Effects of Erosion

4.1. On Site

The implications of soil erosion extend beyond the removal of valuable topsoil. Vegetation growth and the effectiveness of cover are directly affected through natural nutrient loss. Seeds and plants can be disturbed or completely removed from the eroded site.

Soil quality, structure, stability and texture can be affected by the loss of soil. The breakdown of aggregates and the removal of smaller particles or entire layers of soil or organic matter weaken the soil structure and even change the texture. The textural changes can affect the water holding capacity of the soil, making it more susceptible to extreme conditions. When large quantities of topsoil have been removed, the physical properties of the soil do not favor the retention of water, hence large quantities of water runoff rendering the soil infertile.

4.2. Off-Site

Off-site impacts of soil erosion are not always as apparent as the on-site effects. Eroded soil, deposited down slopes can inhibit or delay the emergence of seeds and bury small seedlings (Ritter 2012). Sediments deposition may lead to road damage. Sediments which reach streams or water courses can accelerate bank erosion, clog drainage ditches and stream channels, silt reservoirs, cover fish spawning grounds and reduce dam water quality. The fact that rural population growth rates have been higher than the expansion in area under cultivation suggests that pressures are greater than generally thought, especially on the best lands.

5. Techniques for Assessing Soil Erosion

Erosion is a major soil degradation problem confronting both land and water resource management. This has led to the development of a wide variety of techniques for assessing risk across a wide range of scales (Morgan, 1995, Vrieling, 2006; King et al., 2005).

Principles and components of the Revised Universal Soil Loss Equation are applied because the model combines sufficient simplicity for application on a national scale with a comprehensive incorporation of the main soil-erosion factors (Roux et al., 2008). The model groups the influences on erosion into five categories, namely climate, soil profile, relief, vegetation and land use, and land-management practices;

$A = R.K.L.S.C.P$ *(Source:* Renard et al., 1994)

Where:

A is the spatial average soil loss in t/ha·yr

R is the rainfall runoff erosivity factor in MJ.mm/ha·h·yr

K is the soil erodibility factor in t/ha per unit R

L is the slope length factor

S is the steepness factor
C is the cover management factor
P is the support practice factor

Soil erodibility is an estimate of the ability of the soil to resist erosion, based on the physical characteristics of each soil (Le Roux et al., 2008). Generally, soils with faster infiltration rates, higher levels of organic matter and improved soil structure have greater resistance to erosion. Sand, sandy loam and very sandy soil like silty soils are easily eroded.

Slope-scale measurements involve field rainfall simulation studies and the use of delineated runoff plots and provide valuable data on erosion rates of different crop covers and soil types (McPhee et al., 1983' Stern, 1990; Rapp, 1998). The disadvantage of this method is that it applies only to one or a few hillslopes and cannot be directly extrapolated to evaluate and monitor erosion for a whole catchment (Sivapalan, 2003). Slope-scale measurements include field rainfall simulation studies and the use of delineated runoff plots, (Stern, 1990; Rapp, 1998; McPhee et al., 1983), which provide valuable data on erosion rates of different crop covers and soil types.

The other methods which are becoming increasingly important have been designed to analyze and interpret broader spatial scales. Geographic Information Systems (GIS) and remote-sensing techniques, as well as soil erosion models applied within a spatial context play an important role at the regional scale. The advent of recent developments in the application of GIS and remote-sensing technology offer considerable potential for meeting these requirements. Remote-sensing techniques using aerial photographs and satellite remote-sensing data have greatly increased the capacity to record and monitor land degradation at the regional level. Important sensor development has taken place through airborne systems, including photogrammetric methods using stereo images, (Flugel et al., 2003; Kakembo, 1997) synthetic aperture radar interferometry, (Hochschild and Herold, 2001), and hyperspectral remote sensing (Vrieling, 2006). The disadvantage of using airborne systems is that they are useful in the direct identification of erosion; they are not feasible for monitoring erosion at a national scale, for which satellite imagery is better adapted.

Spatial modeling/analysis differentiate between classes of models usually rests on the level of complexity used to represent the soil erosion processes, and on the spatial and temporal resolution of the model. Models fall into three main categories: empirical, conceptual, and physically based models (Merritt et al., 2003). The best-known and widely implemented empirical models for estimating soil loss at the regional scale are USLE (*Universal Soil Loss Equation)* developed in the 1970s by the US Department of Agriculture (USDA), and its upgraded version RUSLE (*Revised Universal Soil Loss Equation*). Although devised for application to small hill slopes, RUSLE and its derivatives have been incorporated into many regional-scale erosion studies across the globe.

Additional information, in particular changes in soil structure resulting from agricultural activities, greatly improves the quality of results. However, complex models tend to be restricted to research catchments and are prohibitive in terms of the time required for implementation on a regional basis as required by government policies (Le Roux, 2007). Certain methods can be implemented in areas with limited data and are particularly useful as a first step in identifying sources of sediment.

Soil erosion modeling suffers from a range of problems including data variability, over-parameterisation, unrealistic input requirements and unsuitability of model assumptions or misleading parameter values in local context and lack of verification data. Despite these limitations, soil erosion models have been modified and applied to regional scales for scenario analysis, and to make objective comparisons that are important for targeting of research and soil conservation efforts in South Africa.

6. Modern Soil Conservation Methods

Most African countries are in the transition phase between communal and individual ownership. Management strategies are embedded in social, political and economic structures that vary and change over time (Palmer 1997). Soil conservation policies were "diversionary" in that they failed to address social pressures compelling resource poor farmers to mismanage natural resources (Bassett, 1993).

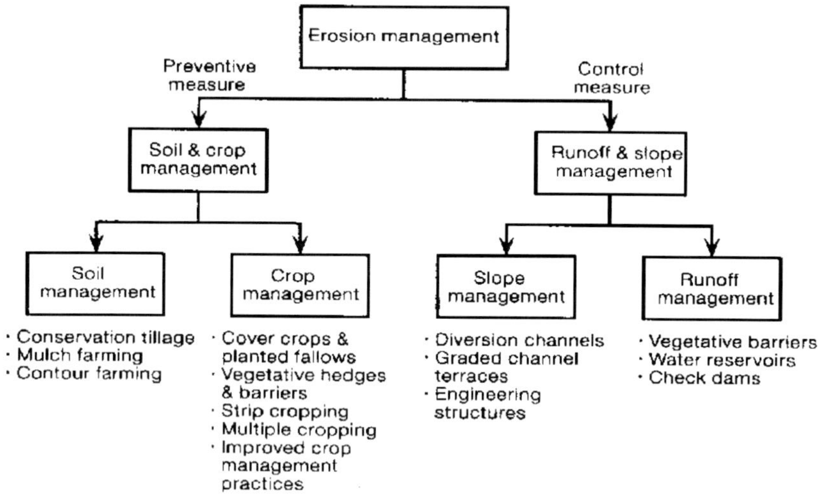

Figure 5. Soil conservation methods (Lal, 1986; Fernandes et al., 1997; Monnier, 1965, Wilson, 1982; Paustian et al., 1997).

The common problem of soil erosion in rural areas in sub Saharan Africa is associated with overstocking, poor agricultural techniques, the fragmentation of landholdings and high population growth (Lana, 1992). The measures used can either be preventive or remedial. Preventive practices minimize the chances of soil degradation occurring or the magnitude or severity of the damage when the degradation manifests. As most countries strive to nurture a class of prosperous African farmers during the period of decolonization, democracy and black empowerment in their nations, so does soil conservation according to Miller (2011) involves methods of reducing soil erosion and restoring soil fertility as indicated in Figure 5.

6.1. Strip Cropping

Strip cropping is used to control both water and wind caused erosion. Strip cropping can be applied in contours or terraces. Strip cropping provides bands of heavier vegetation alternated with row-cropped bands. This method can be used effectively by also crop rotating plants (Camp and Heath -Camp, 2015).

6.2. Crop Rotation Practice

Crop rotation is a traditional soil conservation method where a series of different crops are planted one after the other in the same soil area. This method is used greatly in organic farming. It is done to prevent the accumulation of pathogens which occur if the same plants are grown in the soil; it improves water absorption and also prevents depletion of nutrients.

6.3. Contour Ploughing

Contour farming involves plowing and planting crops in rows across the slope of the land Contour farming or plowing is used by farmers, wherein they plough across a slope and follow the elevation contour lines rather than up and down. This helps to withhold water, reduce water runoff and velocity thus preventing soil erosion by allowing water to slowly penetrate the soil (http://www.buzzle.com).

6.4. Trash Lines and Mulching: Farmers' Innovations

Trash lines and banana mulching are used in Uganda. This is a unique method of controlling soil erosion and improving soil organic matter content, increasing its soil fertility as well as its structure and tilt (Camp and Daugherty, 1997). Trash lines are built every year as a basis for more permanent terrace bunds, the trash lines in south-west Uganda were built along the contour, are kept in place for two or three seasons and then dug into the soil as compost. These trash lines are more than structures to slow down runoff or stop erosion; they also function as mobile compost strips to improve soil fertility. Research results have shown that the use of this method increases organic matter contents, water retention and moisture in the soil above and around the lines Disadvantages mentioned by farmers include alternative uses of the trash and the fact that trash lines tend to hide rats and can act as reservoirs for weed seeds. The results from on-farm research indicated that the mulch application rate had significant effect on yields (Camp and Daugherty, 1997). Mulching with banana trash gave the best returns to labor, as collecting and carrying other mulches requires more labor (Critchley, 1998).

6.5. Vegetative Barriers

Vegetation along roads may allow water to infiltrate and reduce surface runoff. These water ways are usually saucer shaped, trapezoidal or v- bottom (Department of Agriculture, 1999) and are very effective in reducing surface runoff in the Durban city region.

Figure 6a. Plants allow surface infiltration in tarred parking areas. b. Woody shrubs planted along roads to reduce water runoff.

6.6. Re-Vegetation (Vetiver Grass)

Re-vegetation is an effective and cost saving strategy used to prevent soil erosion. Plants help to stabilize the properties of soil and trees act as a wind barrier and prevent soil from being blown away. The use of Vetiver grass (Utshani) is also among strategies used for soil conservation. Vetiver grass species, known as *Vetiveria zizanioidies or Chrysopogon zizanioides* are used in commercial sugar cane farms in South Africa, KwaZulu- Natal, (Lötter et al., 2009). The use of the plant began in 1989 as a form of vegetative hedge along contours in areas with annual rainfall exceeding 1000 mm. The grass is often used as phytoremedial ground cover in disused mines to prevent soil erosion (Lötter et al., 2009; WOCAT 2007). Vetiver grass has unique morphological, physiological and ecological characteristics which and is widely known for its effectiveness in erosion and sediment control (Greenfield, 1989) as in Figures 7a and 7b. In addition, it has been found to be highly tolerant to extreme soil conditions including heavy metal contaminations (Truong and Baker, 1998). In mining areas soil disturbance of mine sites inevitably leads to erosion and to the transport of sand, silt and clay particles in storm water. The grass is planted on mine dumps converting the abandoned space into a series of broad, nearly level terraces that run across the land contours. The seeds of the south Indian cultivar of *Vetiveria zizanioides* are sterile and its roots are not stoloniferous. Thus vetiver hedgerows remain where planted and do not compete with native species. In addition, vetiver hedgerows improve the microclimate and help native species to grow and thrive. A good hedge will reduce rainfall and sediment run off by as much as 70% and 90% respectively. A hedgerow will remain in place and the sediment that is spread out behind the hedgerow gradually accumulates to form a long lasting terrace with vetiver protection.

Figure 7a. Terraces using vetiver grass. (Source: http://en.howtopedia.org/wiki) b. Density and length of Vetiver roots. (Source: http://en.howtopedia.org/wiki)

6.7. Stone Packing

Slope stabilization and runoff control where stones as if they are building a wall. The stone wall helps to trap the soil but allows water to pass through it. Soil builds up over time allowing vegetation to grow. Sometimes the installation of straw "logs" along slope contours is done. These terraces encourage moisture retention and improve surface soil stability while plantings of native vegetation become established. Structural and vegetative practices such as straw bales, silt fences, gravel bags, narrow grass strips or buffers can also be used.

Figure 8. Technique used for bank stabilization near Megacity Mall in Umlazi, KwaZulu-Natal South Africa.

6.8. Prevention of Erosion at Point Sources

Indigenous plants should be encouraged to grow along riverbanks to prevent destabilization of banks. In certain instances, 'contaminated' storm water from mines is filtered by channeling it through a wetland. Grass lawns, gravel, sandstones or any other suitable material can also reduce moisture accumulation (Goudie, 1993).

6.9. Hay Bales and Silt Mesh Fences

These are used for short-term silt traps. The underlying principle behind sediment control is to reduce the velocity of storm water. This causes suspended soil particles to settle out. The larger course sand particles settle first, followed by fine sand, silt and then clay.

6.10. Use of Fertilizers as Soil Erosion Remedy

This includes increased organic and inorganic fertilizer application rates with increasing population pressure at smallholder level. However, the quantities applied are often very modest due to problems of input availability and/or high costs. Prevention of soil detachment by the use of cover materials which will immediately after harvest help protect and hold soil has been widely known for their effectiveness. The use of fertilizers as cover materials is an effective method of restoring lost plant nutrients. Organic and inorganic fertilizers are produced from plants or animals and various minerals respectively. Using beneficial soil organisms like earthworms, helps in aeration of soil and makes the macro-nutrients available to plants. Thus, the soil becomes more fertile and porous. The recycling of organic residues found in soil is currently supported by scientists as a means to mitigate environmental hazards resulting from intensive agriculture practices (Ordon˜ez et al., 2006). The organic refuse wastes of plant and animal origin, provides a good source of nutrients to improve soil productivity (Singh et al., 2010; Gopal et al., 2010; Arthur et al., 2007). According to Tejada et al., (2008) and Gutierrez-Miceli et al., (2008) leachate derived from vermicomposting are beneficial and can be used as liquid fertilizer. Vermicompost and vermicompost leachates are effective in soil conditioning and increase most of the mineral elements which are available in varying degrees and influence soil properties (Ansari, 2008). Vermicompost amendment promotes humidification, the physical, chemical and biological properties of soil increased microbial activity and enzyme production (Kale, 1998). There is good evidence that vermicompost which, in turn, increases the aggregate stability of soil promotes growth of plants (Fosgate and Babb, 1972; Edwards et al., 2000). Biodiversity especially in plants are very crucial for both the culture and livelihoods of most African communities. The progressive development on organic farming is a system of farming based on sustainable integral relationship with nature. Crop production techniques promote plant growth and hence surface cover.

Figure 9. Earthworms are involved in organic fertilizer production (Source: http://www.buzzle.com).

6.11. Soil Conservation Farming Equipment

Modern technologies have developed special tillers which break up and loosen subsurface soil without turning over the topsoil, previous crop residues, or any cover vegetation. In no-till farming, special planting machines inject seeds, fertilizers and herbicides into the thin slits made in the unplowed soil and then smooth over the cut. One of the major constraints to agricultural production in the Sahel is the shortage of rainfall, its uneven distribution and the unpredictable onset of the rainy season (Rain Media, 2003). Plowing leaves the soil bare and exposes it to splash, sheet and wind erosion. Furthermore, if ploughed to the same depth continuously, a hard pan may develop through which rain and roots cannot easily penetrate. Conservation tillage technology for smallholder farmers is now being evaluated in several countries in Africa. Farmers in the Chungu district of Tanzania, where the land preparation window is too short to plough by ox, have tripled their area under cotton using ox drawn weeders after planting in holes hand dug by hoe. In many countries, sowing is also done by hand but animal drawn speeders are easily made and used and save labor and seed. Several factory-made speeders are currently used in West Africa Figure 10a and 10b. The challenge is to develop a critical mass of potential users so that workshops become interested in supplying the market. Training and financial support may be required for jigs, technical drawings and early production required for jigs, technical drawings and early production (http://www.new-ag). This method is faster than plowing; it may also help women, who are usually responsible for growing food crops, to plant earlier and, possibly, larger plots with obvious advantages for household food security.

Less labor is required; it is not necessary to wait until the soil is friable. Thus erosion of any type is avoided.

Figure 10. a. Soil conservation farming equipment. b. Animal driven equipment.

6.12. Wind Breaks

The control measure shelter belts, of trees to minimize or reduce the impact of wind action. These helps to reduce wind erosion, help retain soil moisture, supply wood for fuel and become a habitat for various organisms (Miller and Spool man, 2011).

6.13. Transverse Hillside Ditches

This is a channel has a supporting ridge on the lower side constructed across the slope at defined gradient and horizontal or vertical interval with or without a vegetative barrier and the purpose is to control the flow of water by diverting runoff from upland sloping areas to a stable outlet (NRCS-NHCP 2008), Transverse hillside ditches interrupt and transport runoff to allocated wetlands.

6.14. Tecco Mesh

A mat of high-tensile steel mesh is used in areas where clearing and excavation has made it necessary to stabilize steep rocky slopes especially after road construction by using huge mechanical scaling equipment such as excavators, cranes or explosives. The mesh is tightly anchored into the soil and underlying rock to prevent movement. Seeding provides additional stability following the installation of the mesh (Context Sensitive Rock Slope Design Solutions, 2011). Some temporary practices used on construction sites such as a silt fence placed at the base of the slope will not reduce erosion on the slope but will trap some of sediment leaving that slope.

Figure 11. a. Tecco mesh used to stabilize slopes (Source:http://www.th.gov.bc.ca).
b. Excavating using mechanical equipment rock (Source: Context Sensitive Rock Slope Design Solutions, 2011).

6.15. Geotextile

Geotextiles are one of the common measures used for soil conservation and soil erosion. They cover the soil and stabilize areas susceptible to or exposed to soil erosion. They have received worldwide attention, due to the fact that they come in different materials with more interestingly organic materials which are easily found, biodegradable and renewable For example, Lala Palm (*Hyperleme coriacea*) (Paterson, 2014); pineapple (Hill, 1990). Some geotextiles may be synthetic and the mesh size is easily adjusted as conditions dictate, such as being made from plastic and wire. According to Paterson's (2014), research in South Africa was carried out using mats made from Lala Palm (*Hyperleme coriacea*). The mats made from organic material are long-lasting, and have significant effect on reducing runoff, sediment concentration and splash detachment, while promoting re-vegetation on a range of soils (Rickson, 1998, Davies et al., 2006, Smets, 2011, Bhattacharyya et al., 2012).

6.15. Planting Pockets

In developing countries, communities in urban areas are encouraged to use materials such as filter fabric bags and tires to provide structural reinforced pocket areas containing topsoil to help "islands" of plants to become established. This method is also helpful in promoting vegetable growth in slope areas (Figure 13a-c).

Figure 12a, b, and c. Are showing the different options for using synthetic geotextile materials to compact gully erosion (Nsibande, 2015).

6.16. Salinity Management

Excessive collection of salts in the soil has harmful effects on the metabolism of plants. The toxicity effect varies with different plants and different salts (Goudie, 1993). Salinity can lead to death of the vegetation in areas of saline patches and those become focal points of soil erosion either by wind or water. Salt can be removed through improved drainage or by addition of quantities of fresh water to leach the salt out of the soil. The application of water in farming areas can be modified by sprinklers rather than traditional irrigation methods. In addition, salt accumulating plants can be planted and harvested by photo-remedial action (Goudie, 1993).

Figure 13a, b, c. The use of easily accessible structural material to assist plant growth (Source: http://www.th.gov.bc.ca).

7. Case Studies

7.1. Chelmsford Nature Reserve (South Africa)

The mean annual precipitation of Chelmsford Nature Reserve is 801mm. The area is grassland which has changed from being a forest (infested by invasive alien plants) and settlement areas. A study was done in summer from January to June 2012, when the effects and types of soil erosion were very prominent. The reserve is in a summer rainfall region, with rain between November and February, with a peak in January. The reserve undertakes mosaic burning in order to manage the grass- land. Water is the dominant agent causing erosion in South Africa reducing crop production potential, lowering surface water quality and damaging drainage networks. The study in Chelmsford Nature Reserve showed that the poor road design, poor road maintenance and continuous use of roads contributed 50-80% of the sediment that enters dams and streams (Figure 14a-d). Past soil erosion in this study area has been seen to have an effect on a soil's edibility for a number of reasons. Many exposed subsurface soils on eroded sites tend to be more erodible than the original soils were, because of their poorer structure and lower organic matter. Road erosions contribute both fine and course sediments to streams, and filling pools and reducing habitat complexity. The soil sediments moved from roads are sometimes deposited in streams, dams, affecting channel capacity, water quality. Lastly the erosion of roads also affects the riparian habitat.

Figure 14. a. Effect of vehicles on the road. b. Effect of compaction on abandoned track. c. Effect of vehicles on a disused road within the reserve. Compaction of soil.
d. Effect of compaction on an abandoned vehicle track can lead to gully formation.

The quantity of dust emissions and detachment of soil particles from one another in unpaved road varies linearly with the volume of traffic but in this case vehicle numbers were high during the summer season as many recreational activities take place within the reserve. According to Zungu (pers. comm., 2013), sand are added to heavily eroded soils and clay was added to sandy soils to correct crusting which leads to soil erosion. In addition, compaction was corrected by breaking up the hard layer using a tractor and adding organic matter but this depended on the size of the affected land and manpower availability. Stone packing is commonly used as a measure to control erosion in the Chelmsford Nature Reserve. The stones are held in a metal mesh and arranged as a wall. This barrier helps to trap the soil allowing water to pass and allows the growth of vegetation.

Figure 15a. Deforestation exposes soil to erosion (Source: http://www.rippleafrica.org/environment). b. Sedimentation of Luangwa catchment area (Source: http://www.rippleafrica.org/environment). c. Deforestation to expand arable land (Source: http://www.rippleafrica.org/environment). d. Charcoal being produced illegally. (Source: http://www.rippleafrica.org/environment).

7.2. The Luangwa Zambezi Rivers Confluence Catchment Area in Zambia

The socio-economic status of rural dwellers in Zambia is largely influenced by the activities, taking place along the Luangwa and Zambezi rivers. These activities include fishing, reed harvesting, transportation, overgrazing and subsistence farming (Lukana, 2003). Over exploitation of the forest and land resource has resulted in loss of fertile areas for agricultural productivity. Practices such as shifting agriculture are decreasing as the population increases. As a result the fallow period is shortened as the need to provide food for families become a priority. Farm plots close to the game management areas on the western side of the district lead to human-wildlife conflicts. Deforestation around the rivers catchment area has taken its toll on the two major rivers of the country (Figure 15a). The cultivation and growing of food crops along the Luangwa river basin has greatly contributed to the occurrence of dramatic siltation and sedimentation resulting in the Luangwa River changing its course (Figure 15b). The Luangwa/Zambezi river disasters are related in that both were caused by deforestation (Figure 15c). Furthermore, erosion, gully formation and illegal production of charcoal (Figure 15d) exacerbate the problem of siltation, sedimentation and floods. Erosion in some areas has resulted in the loss of soil fertility and a reduction in arable land. Unless resource poor farmers in these areas adopt suitable farming techniques, their futures look bleak.

8. RESEARCH AND DISSEMINATION OF INFORMATION

8.1. The SADC-ICRAF Zambezi Basin Agroforestry Project

This project is funded by the Canadian International Development Agency (CIDA) and has been run by International Centre for Research and Agroforestry (ICRAF) since 1996 (Benites, 2001). The emphasis of the program is on the facilitation of research-extension-farmer linkages in order to achieve a significant impact with agroforestry practices on the livelihoods of rural people. Two practices to increase soil fertility are improved fallows with fast-growing leguminous trees and maize intercropped with *Glyricidia sepium.* In Zimbabwe, a project of supporting farmers' land literacy (Zimbabwe) has been widely enforced to increase awareness on how farmers can deal with soil problems. For many years conventional extension in Zimbabwe emphasized oral communication, assuming most farmers to be illiterate. As years progressed various tools have been developed to stimulate group exploration, discovery and learning. These participatory extension and research tools have been found to be highly effective in enhancing self-analysis, farmers' understanding of land husbandry and the dynamics of the environment. As a result of this capacity building process - which farmers call "the school of trying" (Chikoro che kuturaya) - farmers develop their own integrated land husbandry approaches (Chuma et al., 2000).

CONCLUSION

The increasing rate of human population induced changes is currently the major environmental concerns causing soil degradation in comparison to natural factors. In South Africa, several developmental activities are undertaken with mitigation measures to limit the loss of topsoil especially in urban areas. In the future, the challenge for most African countries with rapidly growing populations is to help communities to develop a more sustainable approach to the environment. For example, before approval of road construction there must be valid geophysical information, risk analysis and philosophical consideration. These risk analyses are known as Environmental Impact Assessments.

The rapid removal of sediment from the landscape and potential deposition affect the soil as a medium for plant growth and also causes diseases. Techniques to conserve and stop persistent problems of land degradation by erosion are being developed, discovered adopted and implemented worldwide. The concept of managing the soil as natural resources by the people for their own benefit is critical in order to realize the full potential of its benefits. The involvement of communities in areas affected by erosion must be encouraged so that they are capacitated with skills to rehabilitate degraded areas and conserve soil.

ACKNOWLEDGMENTS

Financial support was provided by the Mangosuthu University of Technology, Durban, South Africa. Our gratitude to Dr Adeyemi Aremu for his valuable input and editorial assistance.

REFERENCES

Ansari, A. A (2008). Effect of Vermicompost on the Productivity of Potato (*Solanum tuberosum*), Spinach (*Spinacia oleracea*) and Turnip (*Brassica campestris*). © IDOSI Publications, *World Journal of Agricultural Sciences*. 4, 333-336.

Arthur G.D., Stirk, W.A., Novak, O., Hekera, P and van Staden J. (2007). Occurrence of nutrients and plant hormones (cytokinins and IAA) in the water fern Salvinia molesta during growth and composting, *Environmental and Experimental Botany*. 61 137–144

Bassett, T. J. (1993). Introduction: The land question and agricultural transformation in Sub-Saharan Africa. *Land in African agrarian systems*, 3-31

Bhattacharyya, R., Zheng Yi, Li Yongmei, Tang Li, M. Panomtaranichagul, S. Peukrai, Dao Chau Thu, Tran Huu Cuongd, Truong Thi Toand, Jankauskase, B., Jankauskienee, G., Fullena, M.A., Subedia, M., and Booth, C.A. (2012). Effects of biological geotextiles on aboveground biomass production in selected agro-ecosystems. *Field Crops Research*. 126, 23-36.

Benites, J.R. (2001). The SADC-ICRAF Zambezi Basin Agroforestry Project http://www.fao.org/docrep/003/y1730e/y1730e05.htm Natural Resources Management and Environment Department , Conservation agriculture... chapter 4 Enabling community-based projects.

Camp, W.G and Daugherty, T. B. (1997). Managing our natural resources. 3 rd edition. Delmar publishers, Washington. pp 49-119.

Camp, W.C and Heath-Camp, B. (2015) Managing our natural resources. 8th edition. USA: Cengage learning.

Chuma, E., Mombeshora, B.G., Murwira, H.K., and Chikuvire, J (2000). The dynamics of soil fertility management in communalareas of Zimbabwe. In: Nutrients on the Move-Soil Fertility Dynamics in African Farming Systems. Hilhorst, T. and Muchena, F.M., eds. International Institute for Environment and Development, London. pp. 45.

Critchley, W. (1998). From soil conservation to sustainable land management: a new approach in Africa. In: Critchley, W. Versfeld, D. and Mollel, N. (Eds.). Sustainable land management: some signposts for South Africa. University of the North Press, Sovenga, South Africa. pp. 9-31.

Davies, K., Fullen, M.A. and Booth, C.A. (2006). A pilot project on the the potential contributions of Palm mat geotextiles to the soil Conservation. *Earth Surface Processes and Landforms*. 31, 561-569.

De Graaff, J., Kessler, A and Nibbering, J.W. (2011). Agriculture and food security in selected countries in Sub-Saharan Africa: diversity in trends and opportunities. *Food Security*. 3, 195 213.

Department of Agriculture (1999) *Soil erosion*, [Accessed on 27//02/2015].

Department of Environment. (2005). Climate Change The Fiji Islands Response. Fiji's First National. http://unfccc.int/resource/docs/natc/ fjinc1.pdf [Accessed on 27/02/2015].

Dotterweich, M. (2013). The history of human-induced soil erosion: Geomorphic legacies, early descriptions and research, and the development of soil conservation - A global synopsis. *Geomorphology*. 201, 1–34.

Drechsel, P., Gyiele, L., Kunze, D and, Cofie, O. (2001). Analysis, Population density, soil nutrient depletion, and economic growth in sub-Saharan Africa. *Ecological Economics*. 38, 251–258.

Edwards, L., J.R. Burney J.R., Richter G and MacRae,AH. (2000). Evaluation of compost and straw mulching on soil-loss characteristics in erosion plots of potatoes in Prince Edward Island, Canada. *Ecosystem and Environment*. 81: 217-222.

FAO (Food and Agriculture Organisation of the United Nations). (2001). Soil fertility management in support of food security in sub-Saharan Africa. ftp://ftp.fao.org/agl/agll/docs/foodsec.pdf [Accessed on 26/02/2015].

Fernandes, E.C.M., Motavalli, M.P., Castilla, C and Mukukumbira, L. (1997). Management control of soil organic matter dynamics in tropical land-use systems, *Geoderma*. 79, 49-67.

Flügel, W., Märker, M., Moretti S., Rodolfi,G and Sidrochuk, A. (2003). Integrating geographical information systems, remote sensing, ground thruthing and modelling approaches for regional erosion classification of semi-arid catchments in South Africa. *Hydrological Process*. 17, 929–942.

Fosgate, O.T and Babb, M.R. (1972). Biodegradation of animal waste by Lumbricus terrestris. *Journal of Dairy Science*. 55, 870-872.

Garland G.G., Hoffman, T and Todd S. (2000). Soil degradation. In A national review of land degradation in South Africa, unpublished report, eds T. Hoffman, S. Todd, Z. Ntshona, and S. Turner, pp. 69–107. South African National Biodiversity Institute, Pretoria. Online at: http://www.nbi.ac.za/ landdeg General Information About the Environment in Malawi and Deforestation in Africa [Accessed on 14/12/2014].

Giller, K.E., Witter, E and Mcgrath, S.P. (1998). Toxicity of heavy metals to microorganisms and microbial processes in agricultural soils. Soil Biology. *Biochemistry*. 30, 1389-1414.

Goudie, A. (1993). The human impact on natural environment. 4th edition. Blackwell Publishers. Pp 39.

Greenfield, J.C. 1989. ASTAG Tech. Papers. The World Bank, Washington D.C.

Gutierrez-Miceli, F.A., Gracia-Gomez, R.C., Rincon, R.R., Abud-Archila, M., MariaAngela, O.L., Gullin-Cruz, M.J and Dendooven, L. (2008). Formulation of liquid fertilizer for sorghum (Sorghum bicolour (L.) Moench) using vermicompost leachate. *Bioresource Technology*. 99, 6174–6180.

Hendrix, C.S and Glaser, S.M. (2007). Trends and triggers: Climate, climate change and civil conflict in Sub-Saharan Africa. *Political Geography*. 26, 695-715.

Hill, B.I. (1990). Cut losses with "Plok" on new pineapple plantings. Dohne agricultural institute, stutterheim, Eastern Cape. *Donhe Agricultural Bulletin*. 12, 20-22.

Hochschild, V and Herold, M. (2001). Hydrological analysis of high resolution multifrequent, multipolarimetric and interferometric airborne SAR data. In *Remote Sensing and Hydrology 2000,* International Association of Hydrological Sciences Publication No. 267. Owe, M., Brubaker, K., Ritchie, J. and Rango,A. (eds). International Association of Hydrological Sciences Press, Wallingford, Oxon.Pp. 349–353.

Kakembo V. (1997). A reconstruction of the history of land degradation in relation to land use change and land tenure in Peddie district, former Ciskei. Ph.D. thesis, Rhodes University, Grahamstown.

Kale, R.D. (1998). Earthworm Cinderella of Organic Farming. Prism Book Pvt Ltd, Bangalore, India, pp: 88.

Kicking Horse Canyon projects http://www.th.gov.bc.ca/kickinghorse/khc_environmental_resp.htm. [Accessed on 09/09/2014]

King C., Baghdadi, N., Lemote V and Cerdan O. (2005). The application of remote-sensing data to monitoring and modelling of soil erosion. *Catena.* 62, 79–93.

Knapen, A., Poesen, J., Govers G., Gyssels, G and Nachtergaele J. (2007). Resistance of soils to concentrated flow erosion: a *review. Earth Science Reviews,* 80, 75–109.

Lal, R. (1986). Conservation tillages for sustainable agriculture. Tropic vs Temperate environment. *Advances in Agronomy.* 42, 84-191.

Lal, R. (2003). Soil erosion and the global carbon budget. *Environment International.* 29, 437– 450.

Lal, R. (2009). Soil degradation as a reason for inadequate human nutrition. *Food Security.* 1:45–57.

Lana, A.F. (1992). Crop protection services in Southern Africa. Crop protection, 11. http://ac.els-cdn.com/0261219492901652/1-s2.0-02612194 92901652-main.pdf?_tid=c07ea12a-bcd9-11e4-8fff-00000aacb35f&acdnat=1424860595_be77f83c36c015b133237a6ba31020d4 [Accessed on 17/02/2015].

Le Roux, J.J., Morgentha, T.L., Malherbe, J., Pretorius, D.J and Sumner, P.D. (2008). Water erosion prediction at a national scale for South Africa. Available on website http://www.wrc.org.zaISSN 0378-4738 = Water SA Vol. 34 No. 3 July 2008 ISSN 1816-7950 = Water SA [Accessed on 28/02/2015].

Le Roux, J.J., Newby, T.S and Sumner, P.D. (2007). Monitoring soil erosion in South Africa at a regional scale: review and recommendations. *South African Journal of Science.* 103, 329-335.

Lötter, L., Stronkhorst L.D and Smith, H.J. (2009). Report: Sustainable Land Management Practices of South Africa. Report Number: GW/A/2009/102. Food and Agriculture Organization of the United Nations. https://www.wocat.net/fileadmin/user_upload/documents/Books/Best_Practices_South_AFrica.pdf [Accessed on 26/02/2015].

McPhee P.J., Smithen A.A., Venter C.J., Hartmann M.O and Crosby C.T. (1983). *The South African Rain Simulator Program for assessing soil loss and run-off.* Technical report 119. *Department of Environmental Affairs*, Pretoria.

Merritt, W.S., Letcher, R.A and Jakeman, A.J. (2003). A review of erosion and sediment transport models. *Environmental Modelling and Software.* 18, 761–799.

Miller, G. and Spoolman, S. (2011). Living in the Environment: Principles, Connections, and Solutions. Belmont: Cengage learning.

Monnier, G. (1965) Action des matihes organiques sur la stabilitt structurale des sols. *Annales Agronomiques, 16,* 327-400.

Morgan R.P.C. (1995). Soil Erosion and Conservation, 2nd edition. Longman, London. Pg 26-98.

National Department of Agriculture in cooperation with Directorate Agricultural Land and Resources Management. 1999. Soil erosion. Land care South Africa. http://www.nda.agric.za/docs/erosion/erosion.htm [Accessed on 20/11/2014].

NRCS, NHCP (2008). Natural resources conservation service conservation practice and standard. Hillside Ditch Code 423. http://www.nrcs.usda.gov/ Internet/FSE_DOCUMENTS/nrcs143_025928.pdf [Accessed on 29/03/ 2015].

Nsibande, T. (2015). Email communication. Golden Gate National Park. Freestate. *Project manager of erosional measure*.072 724 9905.

Ordon˜ez, C., Tejada, M., Benitez, C and, Gonzalez, J.L. (2006). Characterization of a phosphorus–potassium solution obtained during a protein concentrate process from sunflower flour. Application on rye-grass. *Bioresource. Technology*. 97, 522–528.

Palmer, R.H (1997). Contested Lands in Southern and Eastern Africa: A Literature Survey UK: Oxfam.

Pater, S and Mcreynolds, K. (2005). Watershed Soils https://cals. arizona.edu/watershedsteward/resources/docs/guide/(6)Soils.pdf [Accessed on 28/02/2015].

Paterson D.G. (2014). The use of Palm Leaf Muls in Soil Erosion control. PhD degree (soil science) in the faculty of Natural and Agricultural Sciences, University of Pretoria.

Paustian, K., Andrean, O., Janzen, H.H., Lal, R., Smith, P., Tian, G., Tiessen, H., Van Noordwijk, M. and Woomer, P.L. (1997). Agricutultural soils as a sink to mitigate CO_2 emissions. *Soil Use and Management*. 13, 230-244.

Pretorius D.J. (1998). The development of land degradation monitoring and auditing techniques with the aid of remote sensing and GIS technology. ISCW report no. GW/A/98/27. National Department of Agriculture, Directorate Land and Resources Management, Pretoria.

Rapp, I. (1998). Effects of soil properties and experimental conditions on the rill erodibility of selected soils. Ph.D. thesis, University of Pretoria, South Africa.

Rain Media, (2003). Tilling and Timing New Agriculturalist online.9http:// www.new-ag.info/03-4/focuson/focuson6.htm l_[Accessed on 29/02/ 2015].

Regan Jr, J.J. (2011). Land use predictors affecting land disturbance in exurban arivaca, Arizona. University of Arizona. http://arizona. openrepository.com/arizona/bitstream/10150/202972/1/azu_etd_11818_sip1_m.pdf [Accessed on 26/02/2015].

.Renard, K.G. and Freidmund, J.R. (1994). Using monthly precipitation data to estimate the R-factor in the revised USLE. *Journal of Hydrology,* 157, 287–306.

Rickson, R.J. (1988). The use of georextiles in soil erosion control: comparision of performance on two soils. Proceedings of the fith international soil conservation conference, Bangok, 18th, 29th.

Ritter J. (2012) Soil Erosion — Causes and Effects. http://www.omafra. gov.on.ca/english/engineer/facts/12-053.htm. [Accessed on 27/02/2015].

Sivapalan M. (2003). Process complexity at hillslope scale, process simplicity at the watershed scale: is there a connection? *Hydrological Processes*. 17, 1037– 1041.

Smets, T., Poesen, J., Bhattacharyya, R., Fullen, M.A., Subedi, M., Booth, C.A., Kertesz, A., Szalai, Z., Toth, A., Jankauskas, B., Jankauskiene, G., Guerra, A.J.T., Bezerra, J.F.R., Yi, Z., Panomtarachichigul, M., Buhmann, and C., Paterson, D.G., 2011. Evaluation of biological geotextiles for reducing runoff and soil loss under various environmental conditions using laboratory and field plot data. *Land Degradation and Development*. 22, 480–494.

Sommer, R., Bossio, D., Desta, L., Dimes, J., Kihara, J., Koala, S., Mango, N., Rodriguez, D., Thierfelder, C and Winowiecki, L. (2013). Profitable and Sustainable Nutrient Management Systems for East and Southern African Smallholder Farming Systems – Challenges and Opportunities. A synthesis of the Eastern and Southern Africa situation in terms of past experiences, present and future opportunities in promoting nutrients use in Africa. http://repository.cimmyt.org/xmlui/bitstream/handle/10883/4035/ 99011.pdf [Accessed on 26/02/2015].

Stern, R. (1990). Effects of soil properties and chemical ameliorants on seal formation, runoff and erosion. D.Sc. thesis, University of Pretoria, South Africa.

Tejada, M., Gonzalez, J.L., Hernandez, M.T and Gracia, C. (2008). Agricultural use of leachates obtained from two different vermicomposting processes. *Bioresource Technology*. 99, 6228– 6232.

Tilman, D. (1999). Global environmental impacts of agricultural expansion: The need for sustainable and efficient practices. *Proceedings of the National Academy of Sciences*. 96, 5995–6000.

Truong, P. and Baker, D. (1998). Vetiver grass system for environmental protection. Technical. Bulletin, (1998/1).

Vrieling A. (2006). Satellite remote sensing for water erosion assessment: a review. Catena. 65, 2–18.

Wilson, G.F., Lal, R., and Okigbo, B.N. (1982). Effects of cover crops on Soil structure and on yield of subsequent arable crops grown under strip tillage on an eroded Afisol. *Soil Tillage Research. 2,* 233-250.

WOCAT (2007). Case studies. https://www.wocat.net/fileadmin/user _upload/documents/Books/WOOK_PART2.1.pdf [Accessed on 26/02/ 2015].

Zungu, E.M. (2013). Ezemvelo KZNWildlife Conservation Manager. Chelmsford Nature Reserve. Personal interview and information literacy manual.

EDITOR CONTACT INFORMATION

Dr. Manuel Esteban Lucas Borja, PhD
Universidad de Castilla La Mancha
Escuela Técnica Superior de Ingenieros Agrónomos y de Montes
Departamento de Ciencia y Tecnología Agroforestal y Genética
Campus Universitario s/n,
C.P. 02071, Albacete (Spain)
ManuelEsteban.Lucas@uclm.es

INDEX

A

B

C

D

E

F

G

H

I

N

O

P

T